Mechanisms and Machine Science

Volume 58

Series editor

Marco Ceccarelli
LARM: Laboratory of Robotics and Mechatronics
DICeM: University of Cassino and South Latium
Via Di Biasio 43, 03043 Cassino (Fr), Italy
e-mail: ceccarelli@unicas.it

More information about this series at http://www.springer.com/series/8779

Rany Rizk · Mariette Awad
Editors

Mechanism, Machine, Robotics and Mechatronics Sciences

Springer

Editors
Rany Rizk
Faculty of Engineering, CRSI
Lebanese University
Roumieh
Lebanon

Mariette Awad
Department of Electrical
and Computer Engineering
American University of Beirut
Beirut
Lebanon

ISSN 2211-0984 ISSN 2211-0992 (electronic)
Mechanisms and Machine Science
ISBN 978-3-319-89910-7 ISBN 978-3-319-89911-4 (eBook)
https://doi.org/10.1007/978-3-319-89911-4

Library of Congress Control Number: 2018938786

Printed on acid-free paper

This Springer imprint is published by the registered company Springer International Publishing AG part of Springer Nature
The registered company address is: Gewerbestrasse 11, 6330 Cham, Switzerland

Preface

This book presents the proceedings of the first international congress, for the advancement of mechanism machine robotics and mechatronics sciences, held in Beirut during October 2017. This conference was organized by the Faculty of Engineering at the Lebanese University and the Lebanese Federation to Mechanism and Machine Sciences, the official delegation of the IFToMM in Lebanon. The conference was under the responsibility of Dr. Rany Rizk the conference chair. It was a part of the engineering job fair organized by the alumni of the faculty of engineering. During this job fair, a mechatronics competition for the students was organized by the mechatronics club of the faculty of engineering of the Lebanese University. The main goal of the conference was to assemble the largest number of students, researchers, businessmen and industrialists.

The book presents 20 recent research works in six different fields. It covers multiple corners of the machine and robotics sciences. It contains works about mechanical design, control, structural synthesis, vibration study and manufacturing. It is a useful material for mechanical as well as for electrical engineers. Works included in the book present high scientist methodology which makes it interesting for researchers. The scientific committee members have reviewed all the papers, each paper was reviewed by at least two specialized reviewers. The percentage of acceptance was about 75% based on the standard IFToMM evaluation criteria.

Temperature Effects of the Machinery and Innovation in the Engineering

First two chapters present the effect of the temperature on the manufacturing and the machine elements, mainly the cyclic loading due to the temperature fluctuation. Third chapter presents a new method about wireless sensor network. Fourth and fifth chapters are about two innovative devices developed for industrial applications.

Biomedical Engineering

This part contains three chapters in the domain of biomedical engineering, a modelling of biologically inspired structure and two exoskeletons, one for upper and one for lower limb.

Vibrations

Four works about a new absorber and effects of the vibrations of the machine elements are presented.

Mechanisms

Five very interesting works about mechanical design and control of parallel mechanisms and pulley systems are developed. All these works are new to the literature and have a real scientific scoop.

Quadcopter and Mobile Robots

Four interesting works, about different methods of control and localization, are done with high professionalism and scientist methodology.

Finally the book is simultaneously, reach in its contents and short in its volume.

Acknowledgements

This work would not have appeared without the support of several persons and companies. We have to support the Lebanese ministry of industry, mainly the minister who participated personally in the opening. Special thanks to the Faculty of Engineering of the Lebanese University mainly to the Dean who support greatly the conference. We thank also the Lebanese association for the advancement of sciences and the IFToMM.

Of course great thanks for all the sponsors, the Alumni of the faculty of engineering and the AUF (French agency for francophone universities). Finally special thanks for the reader and happy reading.

Roumieh, Lebanon Rany Rizk
Beirut, Lebanon Mariette Awad

Contents

Part I
Temperature Effects of the Machinery

On the Validation of the Proper Generalized Decomposition Method with Finite Element Method: 3D Heat Problem Under Cyclic Loading

A. AL Takash, M. Beringhier, M. Hammoud and J. C. Grandidier

Abstract Polymers are widely used in machine design, biomedical engineering, aircraft, and mechatronics. Certainly, these products are subjected to thermo-mechanical loading. For this reason, dealing with the fatigue criteria for viscous polymers is essential. In fact, this requires a large number of cycles to reach the accommodated cycle. These materials have a thermoviscoelastic behavior that has already been studied within the Finite Element Method (FEM). However, a very large computation time or a non-convergence has been noticed. In this paper, we considered a 3D thermal problem under cyclic load in which three main parts are considered. Firstly, the problem is studied using the Proper Generalized Decomposition method (PGD) where it shows a gain in the computation time compared to the classical FEM with a relative error less than 5%. Secondly, different time scales are taken into consideration: the cyclic time (load) and the characteristic one (material properties), where different relaxation times lead to various responses of the polymer. A link between these times should be established. Finally, the effect of the spatial point position on the evolution of the temperature within the domain is investigated.

Keywords PGD · Thermal · FEM · Computation · Cyclic loading

A. AL Takash · M. Hammoud (✉)
International University of Beirut, Beirut, Lebanon
e-mail: Mohamad.hammoud@liu.edu.lb

A. AL Takash
e-mail: ahmad.takash@liu.edu.lb

A. AL Takash · M. Beringhier · J. C. Grandidier
Institut Pprime, ISAE-ENSMA Poitiers, Poitou, France
e-mail: marianne.beringhier@ensma.fr

J. C. Grandidier
e-mail: jean-claude.grandidier@ensma.fr

R. Rizk and M. Awad (eds.), *Mechanism, Machine, Robotics and Mechatronics Sciences*, Mechanisms and Machine Science 58,
https://doi.org/10.1007/978-3-319-89911-4_1

1 Introduction

Computation mechanics is a branch of science whose goal is to solve an engineering problem in all real life fields. This science is bounded by the curse of dimensionality that refers to the increase of the complexity of the model such as kinetic of complex material, dynamics, vibration, mechatronics [1, 2].

From the numerical point of view, the usage of classical FEM leads to a large computation time due to the incrementation issue. Moreover, in the case of polymer like polyethylene, cyclic behavior does not reach a stabilized state after few fatigue cycles like metals, it has been shown that 200 cycles are required [3, 4]. The behavior of polymer in a large temporal domain presents different characteristic times, thus a fine discretization is required to take into consideration the smallest characteristic time. Therefore, this type of problems involves a large number of cycles and CPU expensive. Model Order Reduction (MOR) can be a solution. Furthermore, MOR are grouped as a posteriori method on which the reduced model is built after inspection of performed solution [5–7]; examples for this category are:

1. Proper Orthogonal Decomposition (POD). It uses some data from which reduced basis is built, but the approximation is not provided. This method has been used in [8]. To clearly understand POD, the reader can refer to [9].
2. Reduced Basis Method (RBM) used to solve parameterized partial differential equation [10].

A priori methods are referred to a hypothesis based on the construction of the reduced model with no inspection of the solution. In this paper, we deal with PGD method which is an example under this category. Many years ago, Ladeveze initiated the use of separated representation in computation mechanics [11, 12]. Ammar and Chinesta developed this method and use it for approximating multidimensional problems. PGD method consists of finding space-time separated approximation of n-order under the form of the summation of the product of two functions, each one corresponds to a specific coordinate [13, 14]. For a complete and recent review of this method, the reader can refer to [15] and the references therein. This method has been used to solve viscoelastic where Hammoud et al. [16] discussed a non-equilibrium state under creep and cyclic loading. Moreover, they show that a combination of PGD and adaptive time step could be efficiently used to predict the viscoelastic behavior. Beringhier et al. [17] solved a thermoviscoelastic coupled problem for one dimensional case. Nguyen et al. [18] used PGD method to solve coupled transient multiphysics problems in 2D with different characteristic times. Ammar et al. [19] studied a transient simulation of mechanical behavior where the characteristic time of the mechanical response is less than the time of interest that is using a very small time step. Also, they used space-time separated representation to simulate an integro-differential model. They mentioned that for the case of viscoelastic behavior (time dependent), the standard incremental simulations are inefficient. Bergheau et al. [20] used the PGD as a space-time integrator of elastoplastic problems and they proved the ability of determination of elastoplastic states resulting from cyclic loadings.

Comte et al. [21] introduced a direct method to solve thermoelastoplastic, where this method belongs to large time increment and usage of the wavelet transform. They proved that there is agreement with the incremental method. Boisse et al. used LArge Time INcrement (LATIN) method to solve elastoplaciticity with three different loadings: hardening law, three cycle loadings and parabolic [22]. Congnard et al. extend LATIN method for several hundred cycles in a single increment for case of viscoplasticity [23].

In light of these observations, this paper presents preliminary study and it will be extended to a three dimensional thermoviscoelastic. The objective of the present study is to overcome the huge time computation we may face with FEM, to link the behavior of material under cyclic load, understand the link between cycle time and characteristic time, and to shed lights on the effect of the spatial position. The paper is organized as follows: modeling of the thermal problem under Neumann boundary condition and problem description, then the mathematical formulation and numerical procedure are presented. Numerical simulations and the associated results are discussed. Finally, the paper is ended up by a conclusion.

2 Modeling

The thermal model is defined in the domain $\Omega_{(x,y,z,t)}$ where a three dimensional transient heat problem under convection is considered. The model geometry is a cube where the spatial boundary is denoted by $\partial\Omega$. The formulation of heat equation is written as follows:

$$\frac{\rho C_p \partial T}{\partial t} - k\Delta T - r = 0 \tag{1}$$

where r refers to the radiation effect. In our case, it is considered null, k is thermal conductivity and it is considered constant since the material is isotropic. C_p refers to specific heat, and ρ is the density. The initial temperature is assumed null. A convection boundary condition is considered $q = h(T - T_\infty)$ where T_∞ is the sink temperature considered as cyclic in our case as shown in Fig. 1.

3 Theoretical Analysis

In this part, the mathematical formulation and the separation form are shown. Let us consider the heat Eq. (1) in three-dimensional cube. The weighted residual form is written as follows:

$$\int_\Omega T^* \rho C_p \frac{\partial T}{\partial t} d\Omega - k \int_\Omega T^* (\frac{\partial^2 T}{\partial^2 x} + \frac{\partial^2 T}{\partial^2 y} + \frac{\partial^2 T}{\partial^2 z}) d\Omega - \int_\Omega T^* r = 0 \tag{2}$$

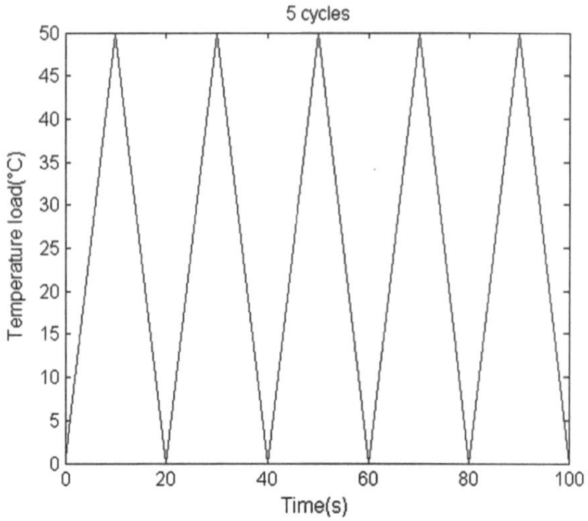

Fig. 1 The first five cycles of the cyclic loading

$$\int_{\Omega} T^{*}\rho C_{p}\frac{\partial T}{\partial t}d\Omega + k\int_{\Omega} T^{*}(\frac{\partial T^{*}}{\partial x}\frac{\partial T}{\partial x} + \frac{\partial T^{*}}{\partial y}\frac{\partial T}{\partial y} + \frac{\partial T^{*}}{\partial z}\frac{\partial T}{\partial z})d\Omega$$

$$-\int_{\partial\Omega} T^{*}(\frac{\partial T}{\partial x} + \frac{\partial T}{\partial y} + \frac{\partial T}{\partial z})d\Omega - \int_{\Omega} T^{*}r = 0 \quad (3)$$

where T is the test field and T^{*} is the virtual field. Let us consider that the boundary condition at the horizontal surfaces of the cube reads:

$-k\frac{\partial T}{\partial n} = h_{horizontal}(T - T_{\infty})$ and at the vertical surfaces $-k\frac{\partial T}{\partial n} = h_{vertical}(T - T_{\infty})$

By taking into account the boundary condition, the equation is written in a separable form.

$$\int_{\Omega} T^{*}\rho C_{p}\frac{\partial T}{\partial t}d\Omega + k\int_{\Omega} [T^{*}(\frac{\partial T^{*}}{\partial x}\frac{\partial T}{\partial x} + \frac{\partial T^{*}}{\partial y}\frac{\partial T}{\partial y} + \frac{\partial T^{*}}{\partial z}\frac{\partial T}{\partial z})d\Omega$$

$$+ \int_{S_{yz}\times\Omega_t} T^{*}(\frac{L}{2}, y, z, t)h_{vertical}(T(\frac{L}{2}, y, z, t) - T_{\infty})dydzdt$$

$$+ \int_{S_{xz}\times\Omega_t} T^{*}(x, \frac{L}{2}, z, t)h_{horizontal}(T(x, \frac{L}{2}, z, t) - T_{\infty})dxdzdt$$

$$+ \int_{S_{xy}\times\Omega_t} T^{*}(x, y, \frac{L}{2}, t)h_{vertical}(T(x, y, \frac{L}{2}, t) - T_{\infty})dxdydt = 0 \quad (4)$$

4 Implementation and Results

The simulation test is a 100 mm long three-dimensional cube subjected to cyclic heat flux in all sides, with null initial condition. Symmetrical boundary conditions are applied and the other surfaces are under Neumann condition. The model encounters a cyclic load with amplitude (50 °C), the time domain denoted by L_t equals 1000*s* and leads to consider 50 cycles (20*s*/cycle). The polymer used is polyethylene having the following properties:

$\rho = 1000$ kg.m^{-3}
$C_p = 1596.66$ J.kg.C^{-1}
$k = 0.45$ W.m^{-1}.C^{-1}
$h_{vertical} = 4.6$ W.m^{-2}K^{-1}
$h_{horizontal} = 6.36$ W.m^{-2}K^{-1}

The characteristic time $\tau_\theta(s)$ is represented by the following equation:

$$\tau_\theta = \frac{\rho C_p}{k} L^2 \tag{6}$$

Remark 1 The finite elements results are obtained using ABAQUS software. We have used automatic discretization in this case. Both simulations were done using the same computer so that the comparison between results is meaningful.

Three different relaxation times (characteristic times) are investigated:

- $\tau_\theta = 0.1$ less than the cycle time $\tau_c = 20$ s.
- $\tau_\theta = 10$ same order as the cycle time $\tau_c = 20$ s.
- $\tau_\theta = 1000$ higher than the cycle time $\tau_c = 20$ s.

To analyze the effect of the characteristic time, let us represent the solution of the temperature using PGD method for three different relaxation times $\tau_\theta = 0.1$, $\tau_\theta = 10$, $\tau_\theta = 1000$ in Figs. 2, 3 and 4 respectively. In fact, the evolution of the temperature depends on the value of the characteristic time with respect to the cycle time. For small τ_θ less than the cyclic time, the stabilized cycle is quickly reached and this can be explained by the fact of the effect of thermal diffusivity where in this case it has a high value. For τ_θ of the same order of the cycle time, the stabilized cycle is attained after certain time. However, when τ_θ is larger than the cycle time, the stabilized cycle is not reached at the end of the simulation due to the low thermal diffusivity. Such responses were witnessed in the 1D viscoelastic case [16].

Comparison of the result to FEM for a fixed spatial point (center of the model) shows that there is an agreement between both methods as shown in Figs. 2, 3 and 4. The PGD method allows predicting the behavior of a polymer under cyclic load with a relative error less than 5%. Furthermore, PGD saves a computational time with the ratio equal to 20 times. The gain ratio is presented in Table 1.

Figure 5 represents the first 5 modes in four dimensions (x, y, z, t) for $\tau_\theta = 10$. We can see that the modes in direction x and z are the same due to the fact that

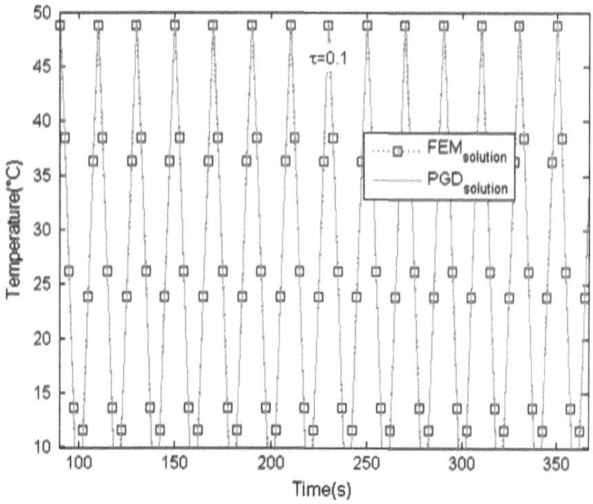

Fig. 2 Validation of the PGD solution for $\tau_\theta = 0.1$ s compared to FEM

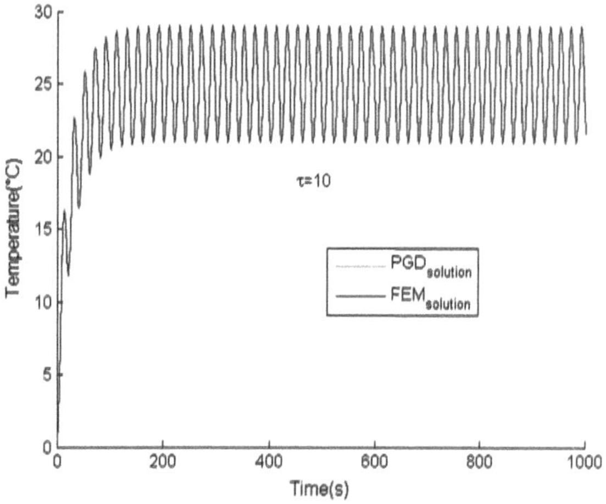

Fig. 3 Validation of the PGD solution for $\tau_\theta = 10$ s compared to FEM

Table 1 Computation time: comparison with respect to FEM

Characteristic time	Gain ratio
$\tau = 0.1$	20
$\tau = 10$	28
$\tau = 1000$	38

the convection coefficient is same in these directions which are not the case of y. However, the modes generated from time domain shows that they are like the load applied.

Figure 6 depicts the evolution of the numerical error and the total computing time on a computer with respect to the number of the modes for $\tau_\theta = 1000$ s. Similarly, for all cases as the number of modes increases the error decreases.

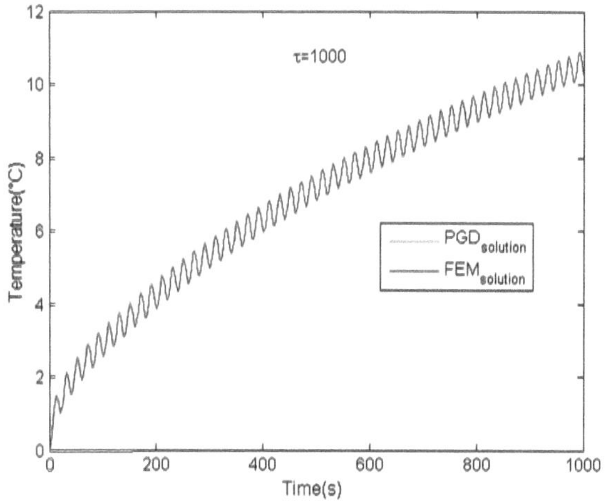

Fig. 4 Validation of the PGD solution for $\tau_\theta = 1000$ s compared to FEM

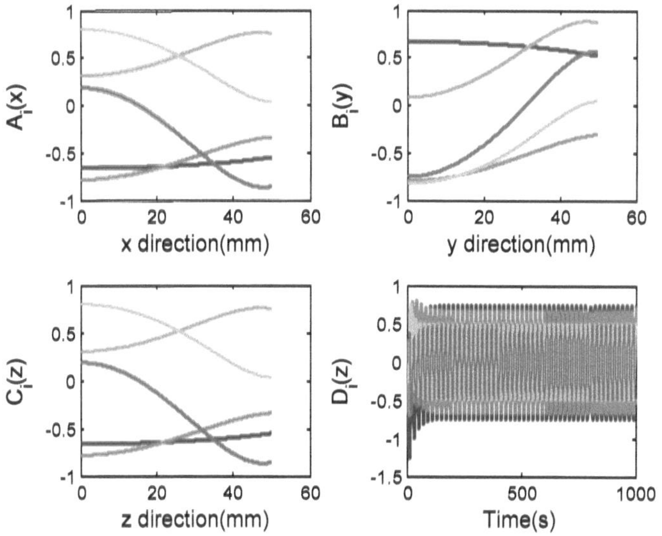

Fig. 5 The first 5 modes in separated form $A_i(x)$, $B_i(y)$, $C_i(z)$, $D_i(t)$

As the spatial position affects the evolution of the temperature, we next analyze whether the spatial position is linked to times scales or not so we used the same model but we applied a cyclic load with a cycle time equal to 100 s and three different characteristic times. Figure 7 shows that regardless of the spatial position, the evolution of temperature is very fast, consequently, the results are the same as in the three different positions. In Fig. 8 we can see for the same order of cycle time and characteristic time the evolution of temperature is approximately the same in all

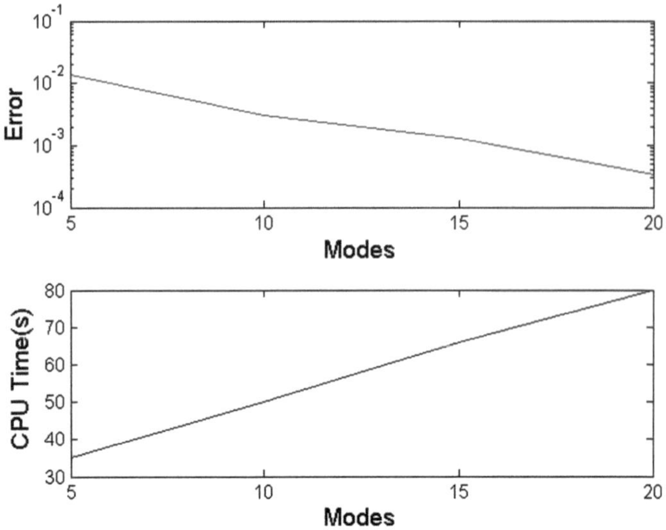

Fig. 6 Evolution of the numerical error and the CPU time with respect to the number of modes for $\tau_\theta = 1000$ s

Fig. 7 Different spatial position with $\tau_\theta = 0.1$ s and cycle time = 100 s

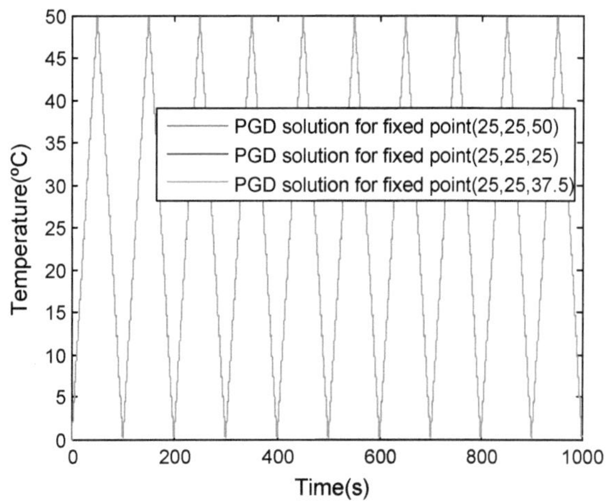

different positions. However, in the case of high relaxation time, Fig. 9 shows that the magnitude and the behavior are affected by the distance from the boundary. Thus, we can link here that the relaxation time affects the evolution of temperature. It can be concluded that when we have small relaxation time the evolution is fast, however in the case of high relaxation time the effect of cyclic load starts to be negligible as we go far from the boundary.

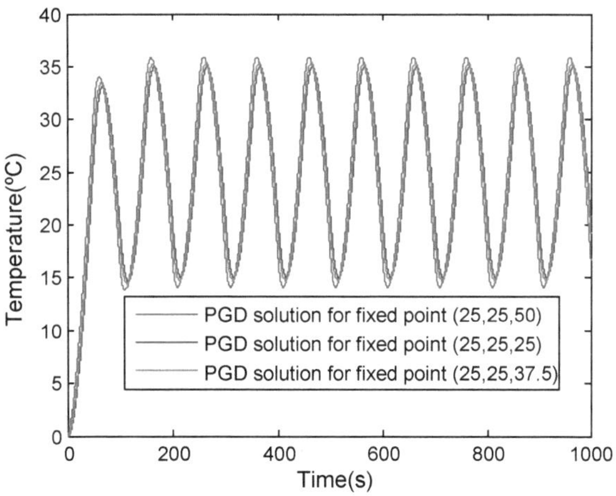

Fig. 8 Different spatial position with $\tau_\theta = 10$ s and cycle time = 100 s

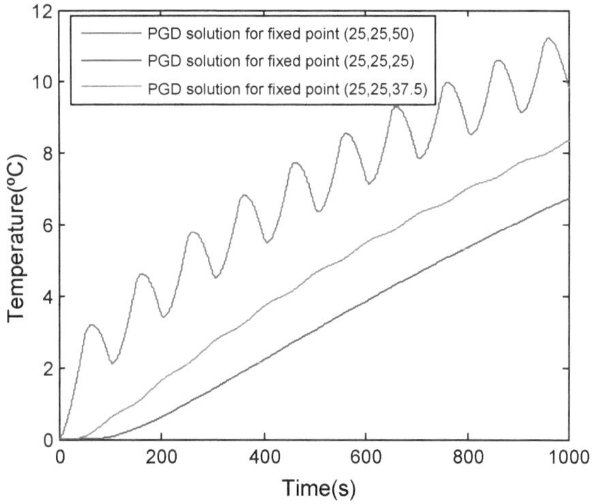

Fig. 9 Different spatial position with $\tau_\theta = 1000$ s and cycle time = 100 s

5 Conclusion

This study is a preliminary study to be extended to three-dimensional thermoviscoelasticity. In this paper, the 3D heat equation has been investigated. The study focused on three parts. A 3D numerical simulation enhancement using PGD was done and the effects of the characteristic times and cycle time has been studied. The PGD allows obtaining a large time saving compared to the FEM. The same responses, as in the 1D viscoelastic case [16], have been obtained. New time basis different than those conventionally used (linear bases) could be suggested to accelerate the method in case of cyclic phenomena. Nonlinearities will be addressed to extend this method to more realistic applications.

References

1. Nguyen STT (2013) Experimental characterization and thermo-mechanical modeling of cyclic behavior of polyethylene, PhD dissertation, ISAE-ENSMA Ecole Nationale Supérieure de Mécanique et d'Aérotechique—Poitiers
2. Bellman RE (2003) Dynamic programming. Courier Dover Publications, New York, republished edition
3. Berrehili A, Nadot Y, Castagnet S, Grandidier J, Dumas C (2010) Multiaxial fatigue criterion for polypropylene automotive applications. Int J Fatigue 32(8):1389–1392
4. Dao KC, Dicken DJ (1987) Fatigue failure mechanisms in polymers. Polymer Eng Sci 27(4):271–276
5. Aguado JV (2015) Advanced strategies for the separated formulation of problems in the Proper Generalized Decomposition framework. PhD thesis, Ecole Centrale de Nantes
6. Benner P (2015) A survey of projection-based model reduction methods for parametric dynamical systems. SIAM Rev 57(4):483–531
7. Chinesta FF, Huerta A, Rozza G, Willcox K (2014) Encyclopedia of computational mechanics
8. Liang Y, Lee H, Lim S, Lin W, Lee K, WU C (2002) Proper orthogonal decomposition and its applications part i: Theory. J Sound Vib. 252(3):527–544
9. Kerschen G, Golinval J-C, Vakakis AF, Bergman LA (2005) The method of proper orthogonal decomposition for dynamical characterization and order reduction of mechanical systems: An overview. Nonlinear Dyn 41(1):147–169
10. Hess MW, Benner P (2013) Fast evaluation of time 2013; Harmonic Maxwell's equations using the reduced basis method. IEEE Trans Microw Theor Techn 61(6):2265–2274
11. Ladevèze P, Passieux J-C, Néron D (2009) The LATIN multiscale computational method and the Proper Generalized Decomposition. Comput Methods Appl Mech Eng 199(21–22):1287–1296 Elsevier
12. Néron D, Ladevèze P (2010) Proper generalized decomposition for multiscale and multiphysics problems. Archives Comput Methods Eng 17(4):351–372
13. Ammar A, Mokdad B, Chinesta F, Keunings R (2006) A new family of solvers for some classes of multidimensional partial differential equations encountered in kinetic theory modeling of complex fluids. J Nonnewton Fluid Mech 139(3):153–176
14. Ammar A, Mokdad B, Chinesta F, Keunings R (2007) A new family of solvers for some classes of multidimensional partial differential equations encountered in kinetic theory modelling of complex fluids: Part Transient simulation using space-time separated representations. J Nonnewton Fluid Mech 144(23):98–121
15. Chinesta F, Ammar A, Leygue A, Keunings R (2011) An overview of the proper generalized decomposition with applications in computational rheology. J Nonnewton Fluid Mech 166(11):578–592
16. Hammoud M, Beringhier M, Grandidier J-C (2014) A reduced simulation applied to the viscoelastic fatigue of polymers. Comptes Rendus Mécanique 342(12):671–691
17. Beringhier M, Gueguen M, Grandidier JC (2010) Solution of strongly coupled multiphysics problems using space-time separated representations application to thermoviscoelasticity. Arch Comput Methods Eng 17(4):393–401
18. Nguyen TL (2012) La décomposition propre généralisée pour la résolution de problèmes multiphysiques transitoires couplés dédiés à la mécanique de matériaux - maillage adaptif et couplage avec la MAN, PhD thesis, Ensma
19. Ammar A, Zghal A, Morel F, Chinesta F (2015) On the space-time separated representation of integral linear viscoelastic models. Comptes Rendus Mécanique 343(4):247–263
20. Bergheau J-M, Zuchiatti S, Roux J-C, Feulvarch E, Tissot S, Perrin G (2016) The proper generalized decomposition as a space time integrator for elastoplastic problems. Comptes Rendus Mécanique 4863(11):753–806
21. Comte F, Maitournam H, Burry P, Mac T (2006) Lan Nguyen A direct method for the solution of evolution problems. Comptes Rendus Mécanique 334:317–322

22. Boisse P, Bussy P, Ladevèze P (1990) A new approach in non-linear mechanics: the large time increment method. Int J Numer Meth Eng 29:647–663
23. Congnard J-Y, Ladevèze P (1993) A large time increment approach for cyclic viscoplasticity. Int J Plast 9:141–157

Qualification and Validation of an in-situ Measurement Method of the Machining Temperature

D. Kara Ali, N. Benhadji Serradj and M. E. A. Ghernaout

Abstract Metal cutting plays an important role in the industry, and as such, the mastery of machining processes has become an issue in the industrial sector. The understanding of the physical phenomena present during manufacture is an important factor in the optimization of a process of production by removal of material. The tool-part interface during a machining process is a place of complex interactions between mechanical, thermal and physicochemical phenomena. The temperature variable which changes according to the cutting parameters directly affects the quality of production. In our contribution, we propose an experimental measurement methodology, which makes it possible to obtain coherent temperature data at the tool-piece interface. This work will focus on two main activities:

- The installation of an instrumentation capable of following the thermal behavior during production. This experimental device is equipped with an infrared camera and an acquisition chain.
- The development of a new device that goes through a qualification stage in which a series of tests will establish the capability of the assembly.

Keywords Machining · Temperature · Infrared radiation

D. Kara Ali
Laboratoire IS2 M, Faculté Technologie, Université of Tlemcen, Tlemcen, Algérie

N. Benhadji Serradj
Laboratoire ETAP, Ecole Supérieure Sciences Appliquées, Tlemcen, Algérie

M. E. A. Ghernaout (✉)
Laboratoire ETAP, Faculté Technologie, Université of Tlemcen, Tlemcen, Algérie
e-mail: mea_ghernaout@mail.univ-tlemcen.dz

R. Rizk and M. Awad (eds.), *Mechanism, Machine, Robotics and Mechatronics Sciences*, Mechanisms and Machine Science 58,
https://doi.org/10.1007/978-3-319-89911-4_2

1 Introduction

The process of material removal by machining is a very complex manufacturing process, using different physical phenomena and their associated sciences: continuous media mechanics, dynamic behavior of materials, chemistry, tribology and thermal. At the heart of the process, two main phenomena are in interaction: a very strong plastic deformation in the shear zones and friction of the chip on the cutting face of the tool. Friction of the flank face can also be added to the newly generated surface. The energy generated is then largely transformed into heat. The modification and the deterioration of the machined surfaces and the cutting tools are mainly related to the phenomenon of generating a large heat in the cutting zone. In order to improve the quality of the surfaces produced and the service life of the cutting tools, it seems necessary to determine the temperatures reached during machining and to understand their generation with a view to minimizing them.

2 Methods of Measuring Temperature

The different methods listed by [1] can be classified into two categories according to the diagram in Fig. 1:

- by conduction
- by thermal radiation

3 Methods of Measuring Cutting Temperatures

The manufacture of mechanical parts by removal of chips obviously generates a friction between the tool and the material. This friction produces a heating increase, increasing the temperature of the tool [2]. The different areas of the tool-workpiece contact during machining are defined as follows (Fig. 2):

Zone of primary shear: It comes from the change of direction of flow of the matter; this zone is the seat of shear and intense shear rates generating a high thermal rise due to dissipation.

Secondary Shear Zone: As a result of the friction of the chip on the tool, this area is both subject to high shear rates and a high temperature rise due to friction.

Tertiary shear zone: In this region, the tool exerts a compressive force on the material which is repressed and separates into two parts; the chip and the machined part. The presence of a chamfer between the face of the section and the flank face at this retaining zone favors the creation of reported edges.

Highly damaging mechanisms such as chemical wear (by diffusion, oxidation, etc.) (Fig. 3) depend directly on the temperature, the measurement of this or its

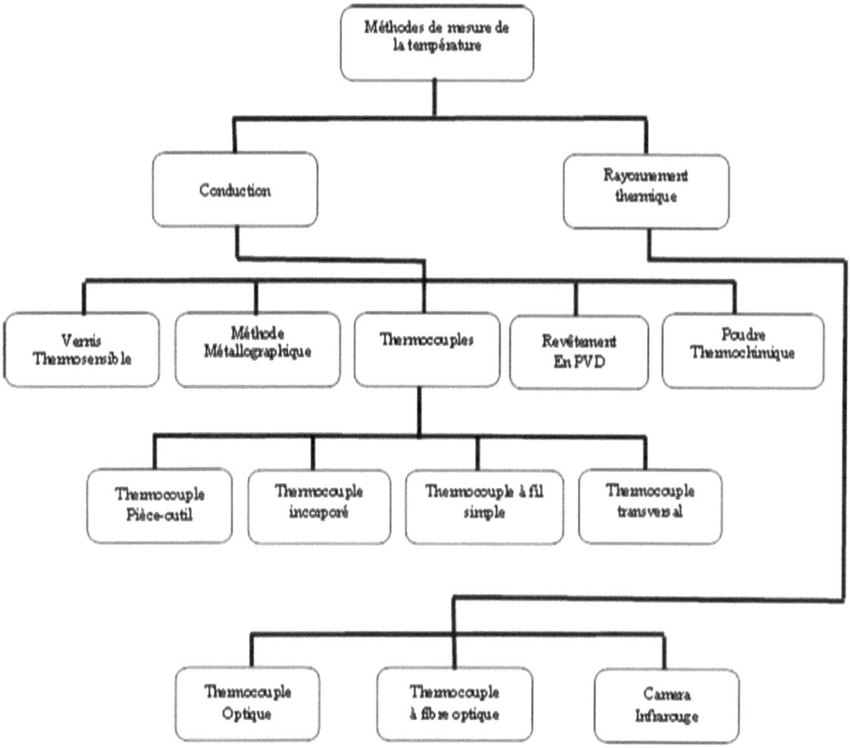

Fig. 1 Different measurement techniques

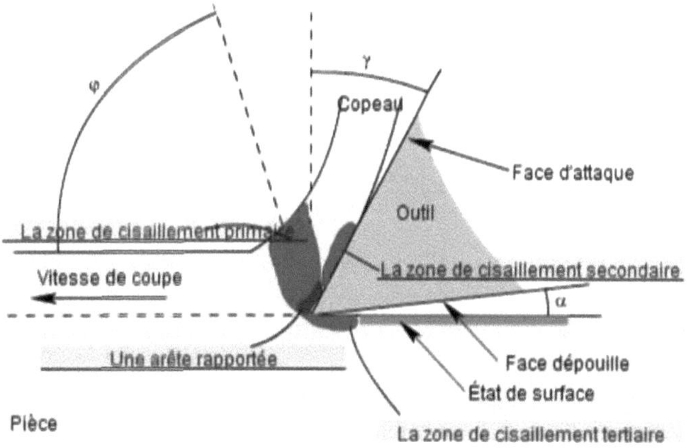

Fig. 2 Chip formation

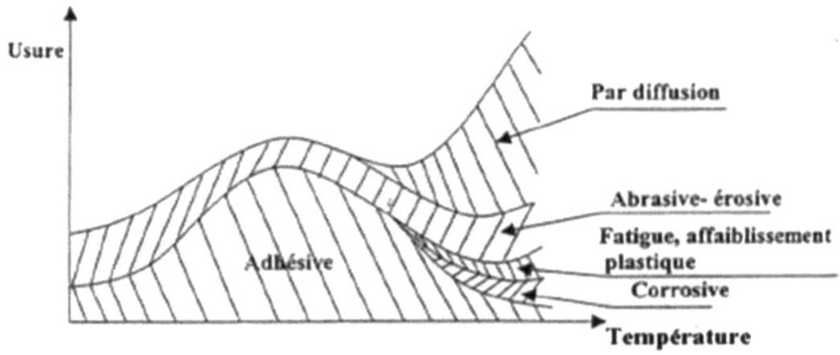

Fig. 3 Influence of thermal and mechanical effects at the workpiece/tool/chip interface [3]

estimation at the chip-tool interface is necessary to the understanding of most physical phenomena of machining it is therefore essential to be able to measure it.

4 Infrared Radiation Measurement

During machining, the stresses applied by the tool on the machined material cause a rise in temperature obviously generated by the friction on the tool-workpiece interface. In this case, infrared thermography is better suited for observing the temperature distribution at the surface of the tool and the generated part. The thermograms thus recorded make it possible to evaluate the radiation emitted by the tool/workpiece pair in a spectral band with a certain acquisition frequency. The temperature variations observed at the surface of the tool and the workpiece are then due to a multitude of phenomena:

- thermal conduction, which tends to standardize the temperature field inside the tool and the workpiece,
- the elastic or plastic deformations induced during machining,
- the dissipation of mechanical energy in the form of heat.

4.1 Infrared Thermography

Infrared thermography is a thermal imaging technique for measuring luminance and temperature. It applies to the entire electromagnetic spectrum. It is a non-contact temperature measurement technique. The diagram below represents the principle of the thermography method: (Fig. 4)

The most common thermographic instrumentation is the so-called "infrared camera" thermal camera. Its function is to transcribe the infrared radiation emitted by an object or system into temperatures that can be used by a treatment system that will display the thermal scenes in the form of thermograms [4]. By design, it has two

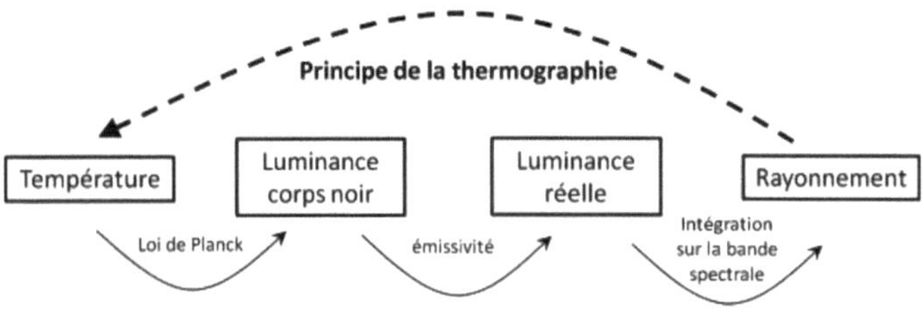

Fig. 4 Principle of thermography

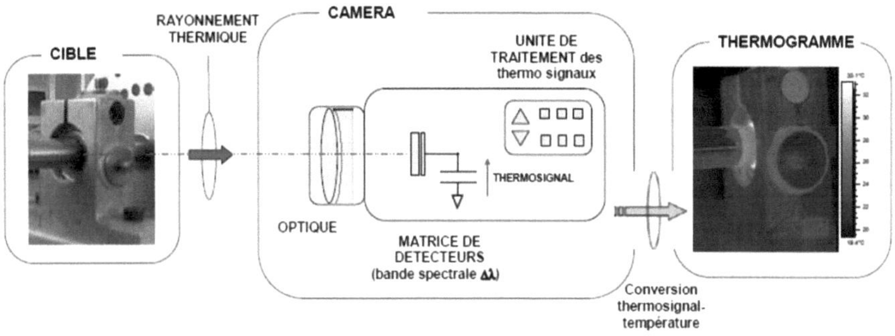

Fig. 5 Radiometric chain of the camera

basic functions: the radiometer and the imager (Fig. 5). As a radiometer, the camera measures a radiation flux and provides, in digital levels, metrologically "clean" thermosignals, images of the total radiation absorbed. As a thinker, the camera multiplexes thermo-signals for the supply of thermograms to digital standards (in principle, 1 measuring point = 1 pixel). It also provides thermosignal-temperature transcription. This transcription is accurate only in one case: the observation of a black body. In other cases, it may not be satisfactory and must then be corrected for the influence parameters of the radiometric measurement (emissivity, environmental temperature, atmospheric temperature, distance, hygrometry, etc.). The analysis is carried out thanks to a powerful software, on an unlimited number of measurement point and by rectangles zones defined by the user. It also offers the ability to capture and save snapshots.

4.2 Theoretical Elements

The increase in temperature due to machining will cause electromagnetic radiation described by Planck's law. Planck's law for a blackbody is the basis of temperature measurement by radiation analysis. This one has a spectral distribution, which depends on the material, its intensity or its luminance ($W.m^{-2}.sr^{-1}$) can be measured and its analysis makes it possible to find the temperature. It is given by equation [1].

$$L_\lambda = \frac{C_1}{\lambda^5\left(e^{\frac{C_2}{\lambda T}}-1\right)} \quad (1)$$

where C_1 and C_2 are constants that are respectively:

$C_1 = 1.191\ 10^{-16}\ W.m^2.sr^{-1}$ and $C_2 = hc/k = 1.43810^{-2} m.K.$

T the absolute temperature in Kelvin degree, λ wavelength in m,

$$\begin{array}{ll} h = 6,626176.10^{-34} J.S & \text{Planck's constant,} \\ k = 1,380662.10^{-23} J.K^{-1} & \text{Boltzmann constant,} \\ c = 2,998.10^8 ms^{-1} & \text{the speed of light.} \end{array}$$

The total flow comprises several components that can disturb the measurement: the own thermal emission (ε) of the target object, a function of its temperature and its emissivity; the flux emitted by the environment (R) and reflected by the surface of the targeted object towards the sensor; and the flux emitted by the ambient atmosphere (τ) separating the object from the sensor. The contribution (τ) is negligible because the space between the sensor and the emitting surface is small (about 10 mm) and the sensor-room assembly is enclosed in an opaque enclosure. Similarly the contribution (R) is negligible given the choice of the wavelength, allowing us to ignore the ambient lighting.

The emissivity ε corresponds to the ratio between the energy radiated by the material at the temperature T and the energy radiated by a black body at the same temperature.

$$\varepsilon = L_{obj}/L_{cn} \quad (2)$$

With L_{obj} = luminance of the measured object

and L_{cn} = luminance of the black body.

Emissivity is a property of surfaces, varies between 0 and 1, depends on the nature of the material, its surface condition, the wavelength and the angle of observation. His knowledge is indispensable for calculating the absolute temperature from the luminous flux signal (Fig. 6). In the case of infrared-opaque nonmetallic materials, it usually approaches 0.9. On the other hand, for metallic materials, it is often weak (<0.5).

The control of a product by infrared thermography, based on the emission of surface radiation, can be done according to two methods:

- passive thermography, which consists of examining the real-time spatial distribution of the apparent temperatures on the surface of an object,
- by active thermography which requires an external energy source to induce a temperature difference between the specimen and any surface or sub-surface defects. The study of the temperature field can thus contribute to the determination of the dynamic stresses generated during a machining operation. This method can in particular highlight thermal phenomena (heating near the cutting edge), represen-

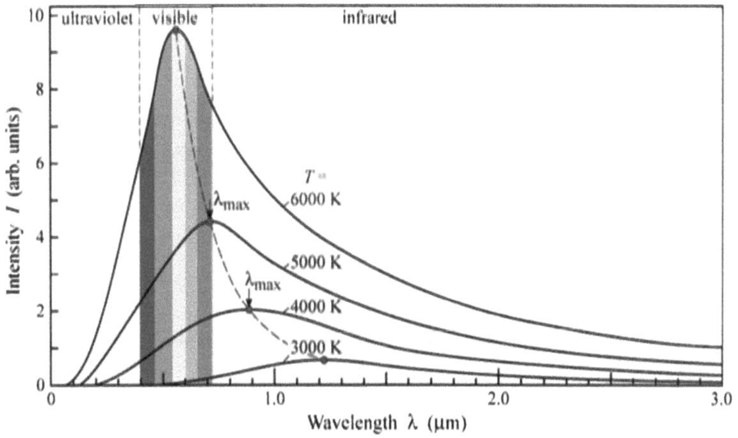

Fig. 6 Spectrum radiation

Table 1 Cutting parameters

Cutting speed	Advance	Depth of the pass	Length of the cut
75 m/min	0.11 mm/tr	0.75 mm	260 mm

tative of mechanical stresses (cutting forces, hardening of the material), around the contact areas tool/workpiece (Fig. 2).

5 Experimental Study

During machining, the temperature on the edge of the cutting edge depends on the cutting parameters. Maximum heat occurs at the tool-room interface. The evaluation of this temperature is very difficult and complicated during a machining operation. The objective of the work is to develop a correlation model for the analysis and prediction of relationships between cutting parameters by measuring temperature at the tool-to-work interface of medium-hard steel during shooting operations. Longitudinal turning is performed on a C20 steel bar 35 mm in diameter on a machine tool installed on a suitable elastic base designed to absorb the vibrations produced by the dynamic forces generated during cutting. The plate used is SNMG square tungsten carbide with a 0° beak angle (Fig. 9).

The cutting parameters used are as follows: (Table 1)

To measure the temperature, we use a **FLIR** infrared camera of type **A325sc**, To measure the temperature, a FLIR infrared camera of type A325sc, with a spectral range of 7.5–13.0 μm, is used which uses a non-cooled and maintenance-free micro bolometer analysis system, with a temperature range of 0 to + 350° C and a precision of ± 2% or 2° C. It can acquire images and data at high speeds up to 60 hertz. It is fixed

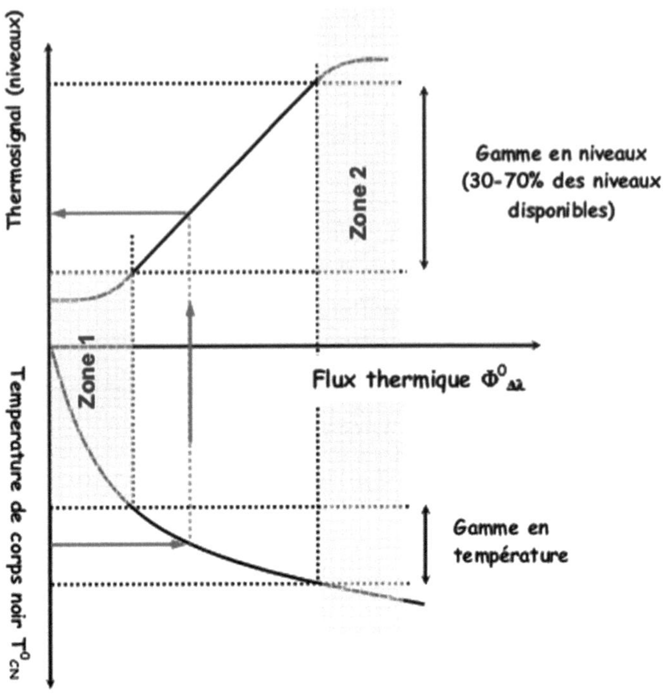

Fig. 7 Zones of the thermosignal

on an adjustable support to give a distance of 30 cm from the tool-piece interaction (Fig. 9). Before its use in experimentation, it must be calibrated. The calibration of the camera uses a black laboratory body whose emission temperature is modified so as to associate the thermosignal given by the camera. The thermosignal is the analogical image of the thermal flux by photoelectric or bolometric conversion. It results from the accumulation of the thermosignal during the integration time. By calibration, a thermal camera assembler looks for linearity of the thermosignal with the flux, as shown in Fig. 7, [5].

The lower and upper limits of linearity are respectively due to insensitivity to low flows.

Zone 1: thermosignal limited by temporal noise and dark signals.
Zone 2: thermosignal limited to saturation with important fluxes or integration time.

Once these limits are established, the measurement dynamic is evaluated by the maximum level gauge on which the thermo-response is the image of the incident flux. It is usually limited to between 30–70% of the available digital levels in order to be able to apply a non-uniformity correction (NUC). For example, in 14-bit conversion, there are 16,384 levels and only the levels between 5000 and 11500 are affected. The sensitivity of the detector (in level/° C) finally makes it possible to evaluate

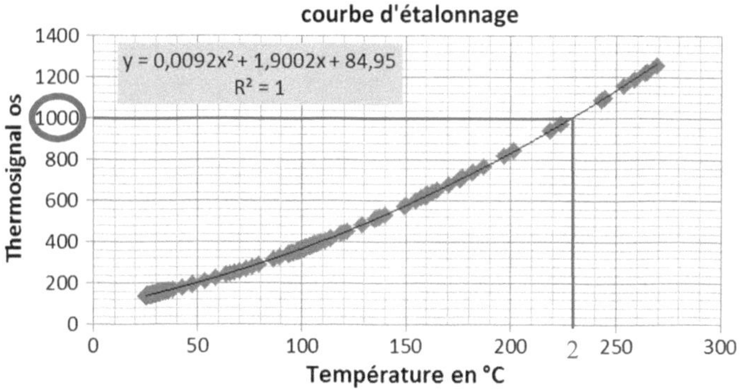

Fig. 8 Calibration curve of the FLIR SC305 camera. (Blue squares: calibration points, line: test function)

Fig. 9 Device for experimentation and installation of the camera

the range in temperature. These choices are essential because they determine the available measurement ranges and the thermal resolution of the cameras.

An example of a calibration curve is given in Fig. 8. The temperature-Thermosignal transcript is a function applied to the test calibration points [6]. The coefficients are stored in the camera for automatic transcription of the thermosignals provided by the camera.

The next photograph (Fig. 9) shows part of the experimental system and the adjustable camera assembly.

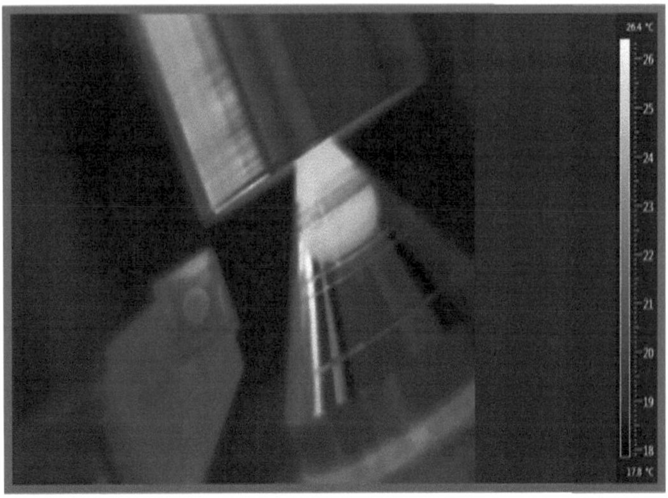

Fig. 10 Thermograms: tool-piece

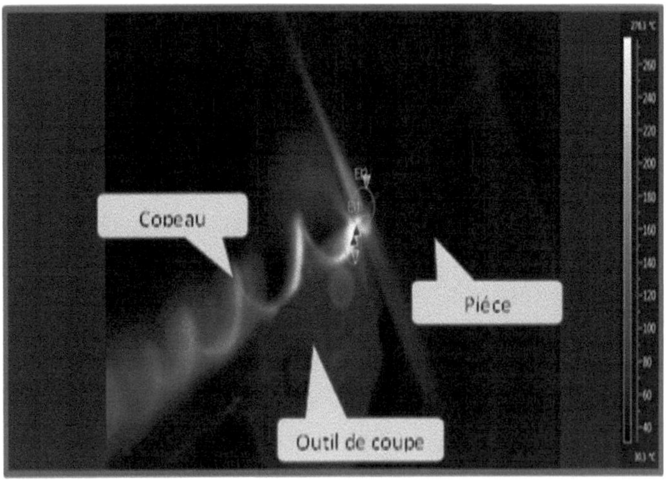

Fig. 11 Thermograms: machining

6 Results and Discussions

Machining performed was recorded. Figures 10 and 11 show the thermograms of the tool-workpiece interface and the turning operation respectively.

The following table groups the results of the experiment (Table 2).

On the wagon length, we chose 10 positions for the temperature reading, using the camera software. These points are taken on the period of stabilization of the machining.

Table 2 Results of experiments

Position	1	2	3	4	5	6	7	8	9	10	T°c moyenne	% (température)
SP1(Piéce)	114,6	117,6	117,5	114,9	117,6	122	118	120,1	113,3	113,4	**116,90**	**23,32**
SP2(Outil)	79,2	89,02	93,3	95,5	96,2	96,3	97	97,3	99,9	98,6	**94,23**	**18,80**
SP3(Copeau)	300,7	298,7	299,5	302	295,5	288,3	299,3	288	291,3	289,2	**295,25**	**58,91**
Σ	495,5	494,5	505,32	510,3	512,4	509,3	506,6	514,3	505,4	504,5	501,2	100

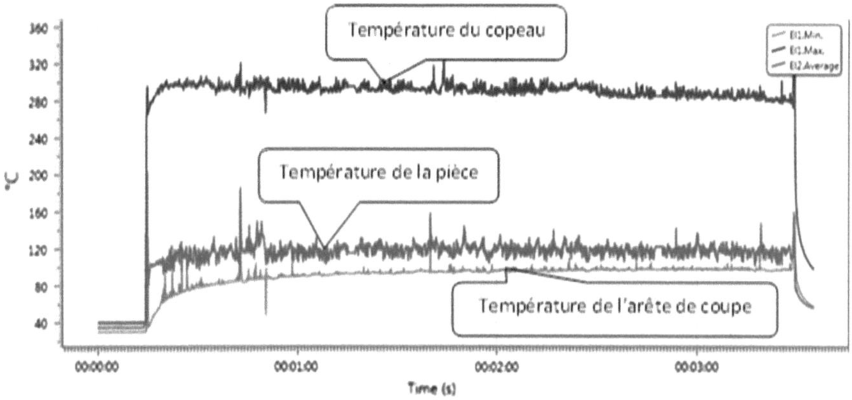

Fig. 12 Temperature variations (Quickplot)

The first row of the table represents the positions, at equal intervals, of the temperature readings belonging to the stabilization period. The second line (SP1) represents the different temperatures of the machined surface of the 10-position part. The third line (SP2) represents the different temperatures of the nozzle of the tool. The fourth line (SP3) represents the chip temperatures. Finally the last line represents the summation of the data by column.

The spectral analysis illustrated in the following figure makes it possible to conclude, on the one hand, that the temperature variations are very small in steady state conditions for invariant section parameters. On the other hand, the heat flow generated during machining is largely evacuated by the chip (Fig. 12).

Using the thermogram data, we plotted the representative curves:

Chip temperatures (in green), machined surface temperatures (in blue) and tool cutting edge temperatures (in red). It can be said that the temperature curves of the machined surface and the cutting edge are almost the same and at very close temperature thresholds (around 100° C). On the other hand, the temperature curve of the chip clearly shows an increase in temperature (in the vicinity of 300° C.) (Fig. 13).

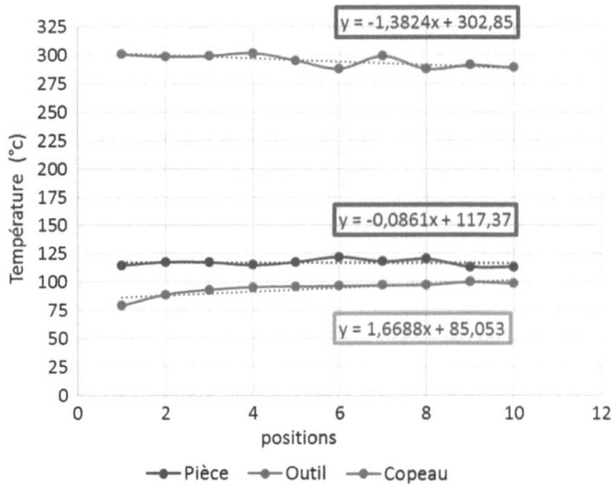

Fig. 13 Temperature variations

7 Conclusions

During machining, the stresses applied by the tool on the machined material cause a rise in temperature of the latter two. This work has shown that a thermography technique is a useful procedure for establishing an experimental methodology to evaluate the cutting temperature during machining operations for different cutting parameters.

The machining conditions we have considered are cutting speed, feed rate and feed depth. The experiment allowed us to determine the temperatures of the machined surface of the workpiece as well as that of the cutting edge.

The experiment confirmed that the chip temperature is still significantly higher than that of the cutting edge and the workpiece.

The major difficulty in taking the machining sequence by the camera is the winding of the chip, hence the need to choose the cutting conditions and to have good protection and fixing of the camera.

Infrared thermography has many advantages such as a non-contact measurement, a measurement without or with very little disturbance between the surface of the object under study and its surrounding environment, a real-time measurement capability, a wide temperature range of use and ability to adapt to any type of materials.

In perspective, we will try to find a correlation between the cutting temperature and the wear of the tool during machining.

References

1. Goyal A, Suresh Dhiman S, Kumar S, Sharma R (2014) A study of experimental temperature measuring techniques used in metal cutting. JJMIE 8(2):82–93 ISSN 1995-6665

2. Chou, S (2005) Thermal modeling for white layer predictions in finish hard turning. Int J Mach Tools Manuf 45, 481–495
3. Opitz H, Konig W (1967) on the wear of cutting tools. In: 8th M.T.D.R. Conference of Montreal, pp 173–189
4. La thermographie infrarouge. 2ème partie. Revue “Technica”. N°441—Nov–Déc. 1984
5. Riou O, Logerais PO, Durastanti JF (2013) Quantitative study of the temperature dependence of normal LWIR apparent emissivity. Infrared Phys Technol 60:244–250
6. Berteleau-Hennart M (2003) une méthode originale et rapide d'étalonnage d'une caméra thermique pour des applications en électronique. Journées d'études de thermographie instrumentale et industrielle. Sénart, 27 et 28 novembre 2003

Part II
Innovation in the Engineering

Simple On-Line Single-View Video Summarization for Machine-to-Machine Wireless Multimedia Sensor Network

Thanaa Jbeily, Iyad Hatem, Mothanna Alkubeily and Yacine Challal

Abstract Wireless Multimedia Sensor Networks (WMSN) engender a wide scope of important Machine-to-Machine video based applications. However, energy consumption in WMSN determines the complete network lifespan. Communication is the greediest energy consumption task which grows exponentially with the increase of transformed data. Even though video data is considered the richest information holder, it is the most energy craving data. In this paper, we propose a new Simple On-line Single-view video summarization mechanism (SOS). This mechanism adds a low-complexity video summarization task to the sensor nodes. Nevertheless, it reduces the transfer load of network through extracting only significant video information. Simulation results confirm the efficiency of our proposition in saving the life of WMSN. Resource consuming reduction reaches as high as 40%.

Keywords Machine-to-Machine Network · Wireless multimedia sensor network
Energy consumption · Video summarization

1 Introduction

Machine-to-Machine (M2M) networks, or Internet-of-Multimedia-Things (IoMT), is the next wave of information technology revolution [1, 2]. Billions of ubiquitous intelligent devices, such as sensors, are connected to communicate with each other in M2M networks. Among different sensors used in M2M networks, wireless multimedia sensors can provide the richest information represented by video data, Fig. 1.

T. Jbeily · I. Hatem (✉) · M. Alkubeily
University of Tishreen, Lattakia, Syria
e-mail: iyadhatem@tishreen.edu.sy

Y. Challal
Ecole Nationale Supérieure D'Informatique, Algiers, Algeria

R. Rizk and M. Awad (eds.), *Mechanism, Machine, Robotics and Mechatronics Sciences*, Mechanisms and Machine Science 58,
https://doi.org/10.1007/978-3-319-89911-4_3

Fig. 1 M2M wireless multimedia sensor network

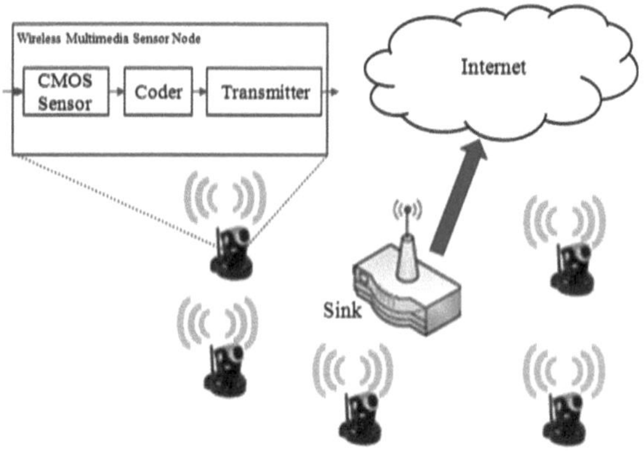

However, Battery lifetime is critical for wireless multimedia sensor networks-WMSN [3, 4], and the energy consumption of wireless multimedia sensor nodes will determine the whole network lifespan. The greediest energy consumption is caused by communication.

Most videos are full of unnecessary redundant data which consumes most of the sensor nodes resources like energy, storage space, bandwidth, etc. [5]. To extract only important information from video, and to reduce the transfer load of WMSN, more and more attention is given to the in-network video processing technology. Recently, video summarization has become an emerging field of interest for medical image processing, sports video analyzing, video indexing for quick browsing, and many other applicable domains [6–8]. The main objective of video summarization is to provide clear analysis of video by removing duplications and extracting key frames (Kf) from the video [9]. Those key frames generate a short synopsis of the original video. However, practical implementation of video summarization is far behind due to complexity of methods. There is a crucial need for implementing simple low-power video summarization mechanisms on video sensor nodes.

In this paper, we suggest a new Simple On-line Single-view video summarization mechanism for WMSN. This mechanism adds a simple summarizing stage to the wireless multimedia sensor node functionality. With this summarization only the most important and interesting contents of videos are produced. When transmitted, these summarized videos saves nodes resources and lets viewers acquire information by watching only few key frames. SOS takes into account simplicity in video summarization process. It does not add huge operating cost on sensor nodes. In fact, it exploits the complementary metal-oxide-semiconductor (CMOS) sensor structure of the nodes to reduce the summarization requirement and to save the quality of experience (QoE) of the end user. Moreover, SOS selects the desired key frames related only to a region of interest of the view.

The remainder of this paper is organized as follows: Sect. 2 covers video summarization for WMSN and the related works. Section 3 proposes our M2M-fitted video summarization mechanism and its details. We evaluate our proposal in Sect. 4 and conclude the paper in Sect. 5.

2 Video Summarization Over Wireless Multimedia Sensor Network

Streaming all video data may waste valuable resources of the WMSN. By introducing the video summarization unit to the sensor nodes, non-informative video data can be removed. Thus, energy saving for both compression and transmission will be achieved. However, most video summarization algorithms depend on feature extraction and comparison to decide the key frames which will later compose synopsis of the original video [6, 7, 10]. Indeed, feature extraction process insert additional new overhead on the limited resources WMSN nodes. Moreover, multi-view based video summarization algorithms impose data exchange between sensors. This exchange is required to be as small as possible. Otherwise, communication overhead will magnify the energy consumption. The preceding challenges make video-fusing summarization techniques in WMSN nodes a very critical task and researches in this area are still investigating it. Authors in [11] consider applying color layout feature comparing multi-view summarization of wireless multimedia sensors to remove redundant contents such that the compression and transmission power can be reduced. They extract key frames based on on-line process which, on the contrary of off-line process, does not require the storage of the whole video in sensor memory in advance. This will remarkably reduce the size of used memory. Nevertheless, the previous study depended mainly on approximate models of sensor nodes and on laboratory measurements of multipurpose boards to evaluate the work, Fig. 2.

However, no previous researches start from the view point of exploiting the origin hardware structure of WMSN nodes to imply video summarization technique.

3 The Proposed Mechanism

Our proposed On-line Single-view video summarization mechanism takes into account the critical constraints of wireless multimedia sensor nodes. It is QoE-aware since it considers the visual and mental perception of the end user. Actually, the distinct features of SOS can be summarized by the followings.

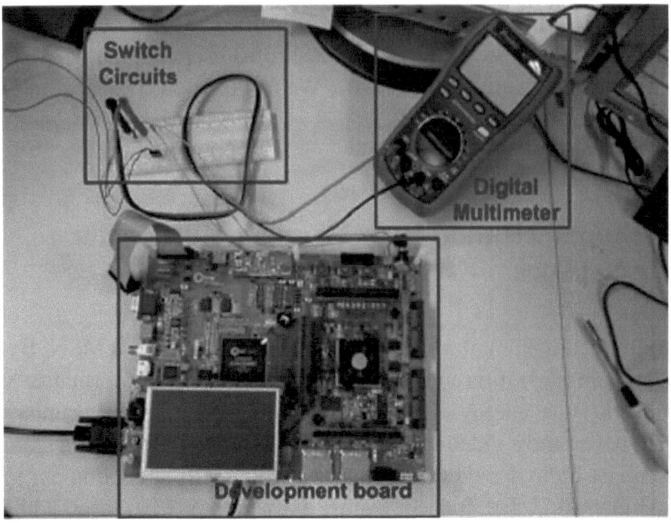

Fig. 2 Measurement environment setup of [11] study

3.1 Spatial Video Data Domain

SOS works in the spatial pixel-based frame domain. In order to extract the key frames, it does not apply any kind of data transformation nor heavy feature extraction tasks.

3.2 Region of Interest

SOS does not examine the entire captured frame to determine if it is a key frame or not. It concerns only with a part of the scene which is considered as Region of Interest (ROI). An example of that desired ROI is a monitored hall with its surrounding entries which is highlighted in a green square, Fig. 3.

3.3 CMOS Sensor Structure and Human Eye Sensitivity

In order to extract the desired key frames for the summarized video, SOS considers only the green color component of the acquired RGB frame and ignores the other red and blue color components. SOS bases in this on the human eye sensitivity characters. The human eye has higher sensitivity to green color compared to red and blue colors [12], as Fig. 4 illustrates. Moreover, SOS exploits the fact that CMOS sensors structure in wireless multimedia sensor nodes are already designed to compact with

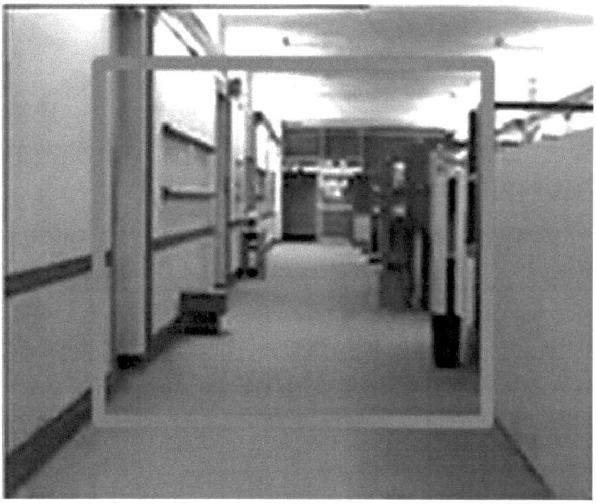

Fig. 3 Region of Interest of the monitoring hall

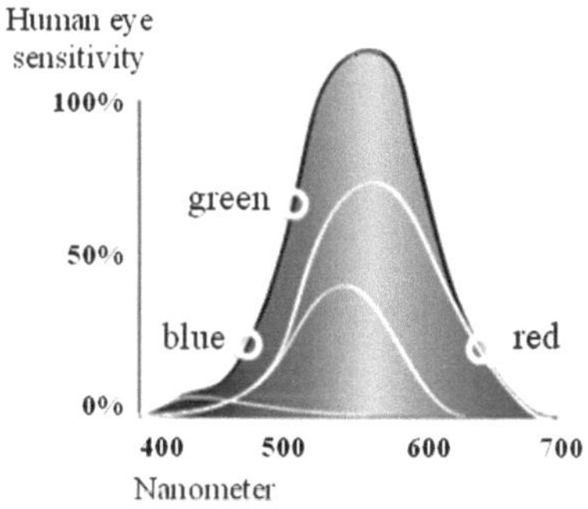

Fig. 4 Human eye sensitivity to color components

Bayer filtering pattern [13]. Each sensory cell in Bayer pattern uses matrix of color filters compatible with human visual perception. Indeed, Bayer distributes matrix as percentages of 50, 25, 25% for green, red, blue color filters respectively [13], Fig. 5.

3.4 On-Line Summarization

SOS adopts on-line mode for video summarization process. This mode does not require selecting and storing the complete video, which is the case in off-line mode. Therefore, SOS decreases the memory requirements. Only one frame buffer is required for comparing every new frame (Nf). This frame is compared, directly as soon as it is captured, with the preceding key frame. SOS takes the decision to concatenate the new frame to the key frame set which will later compose the summarized video. The decision is based on a predetermined threshold value.

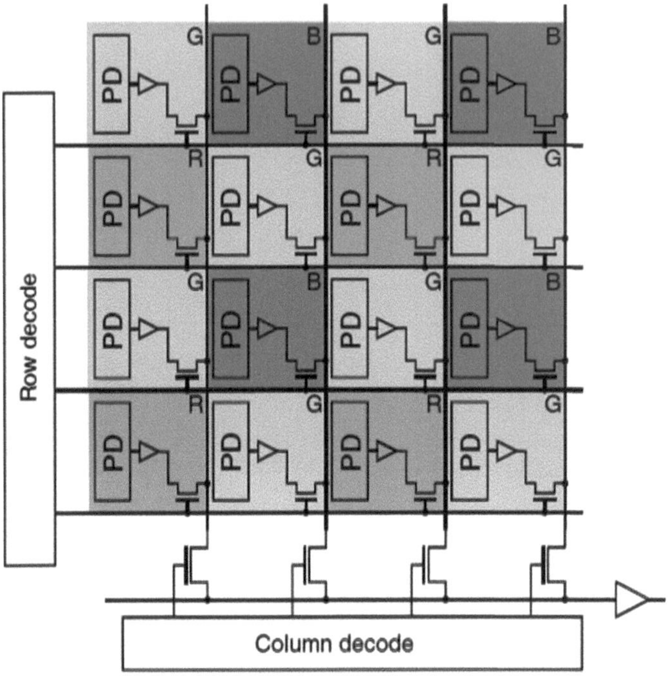

Fig. 5 CMOS sensor arrangement (Bayer pattern filter)

3.5 Single-View Summarization

Wireless multimedia sensor nodes in this study are fixed, not mobile. Each node has fixed directional field of view (FoV) with certain parameters of both depth of view (DoV) and angle of view (AoV). The acquired videos are single-view. Furthermore, sensor nodes do not share their own sensed video information to conserve their resources.

Flow chart in Fig. 6 show the step by step details of our simple low resources-demand video summarization mechanism for WMSN.

4 Implementation and Performance Evaluation

The evaluation of the proposed algorithm, SOS, aims to satisfy the M2M WMSN constraints in video related application. For this purpose, we use a special new OMNet++ framework. It is a Mobile Multimedia Wireless Sensor Network framework, called M3WSN [14]. M3WSN has been designed to support and meet M2M WMSN characters. To our knowledge, it is the first work to exploit M3WSN in evaluating video summarization technique for WMSN.

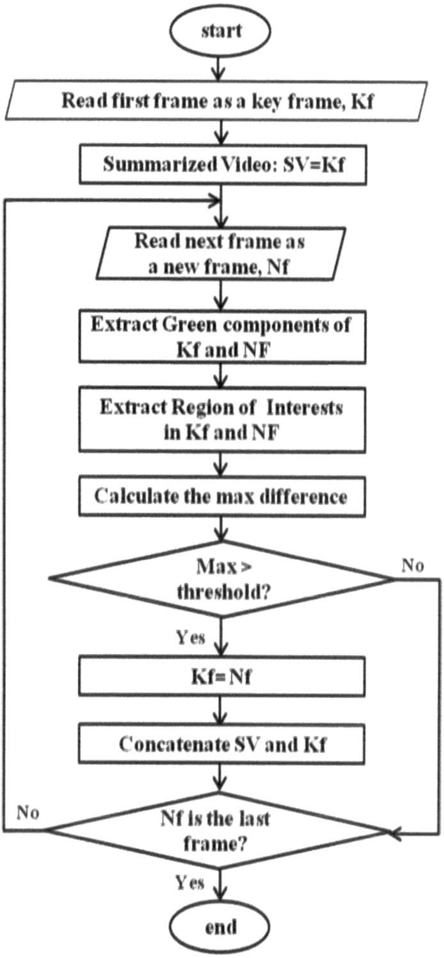

Fig. 6 Flow chart of our proposal, SOS

M3WSN enables the evaluation of new assessment and optimization solutions for fixed and mobile nodes. It efficiently defines a model to find subsets of nodes covering a given area in the network as well as managing the sensing range for each node by FoV. M3MSN framework architecture is illustrated in Fig. 7. The values of design and simulation parameters of our study are given in Table 1.

To evaluate the efficiency of SOS, several performance evaluation metrics are considered.

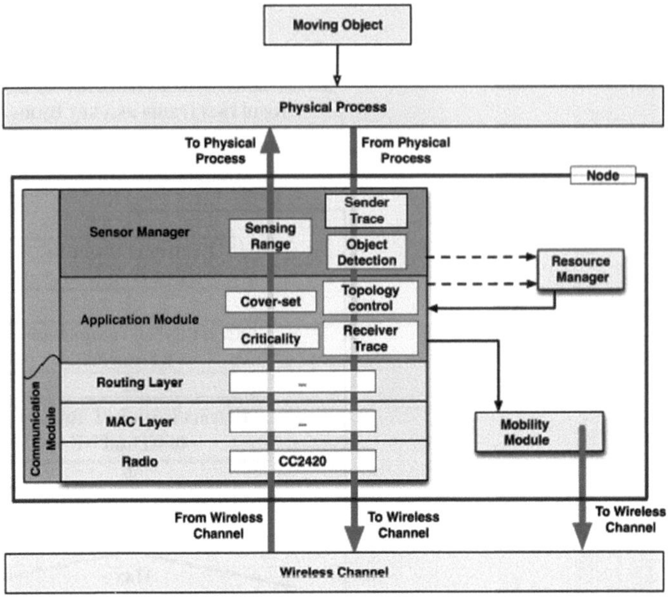

Fig. 7 M2M M3WSN framework architecture [14]

4.1 Number of Key Frames

Simulation results for the obtained number of extracted key frames by applying SOS under different threshold values are shown in Fig. 8. SOS insures reducing the number of frames representing the original video. For example, under threshold value of 100, the summarized video consists only of 177 key frames versus 300 frames in the original video. In this case, SOS offers percentage reduction of 41%. Practically, this reduction will reduce the transmission requirements that, in its order, will be reflected positively on saving the wireless multimedia sensor nodes resources.

4.2 Number of Packets

M3WSN enables setting up the video encoding standards. Video sequence is processed as different kinds of frames (Intra frame-I, Predicted frame-P, and Bidirectional frame-B, in our study). Each frame is encapsulated in a different required number of packets prepared for transmission. The total number of packets is shown in Fig. 9. The results show similar effect of SOS in reducing number of frames. We obtain corresponding reduction in number of packets. The achieved reduction rate is 40.71% for the threshold value of 100.

Table 1 The design and simulation parameters

Design and Simulation parameter	Value
Sink location	(1, 1) mm
Sensor node location	(11, 11)
Sensor manager	Crit Directional Sensor
DoV(m)	15
AoV(rad)	0.314159265359
MAC protocol	Tunable MAC
Access mode	CSMA
Routing protocol	By pass Routing
Path loss model	Lognormal Shadowing model
Radio model	CC2420
Node Initial Energy(J)	4000
Video sequence	Hall
Video encoding	H. 264
Video format	QCIF(144 × 176)
Frame rate	25
Group of picture	31

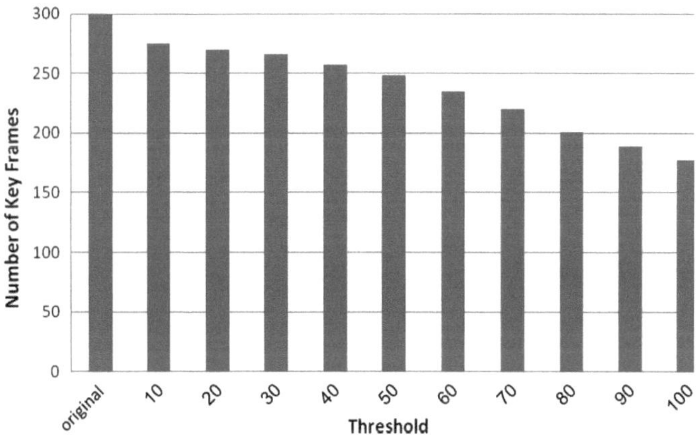

Fig. 8 Number of extracted key frames

4.3 Transmission Energy

Transmission is the greediest energy consumption source in communication. It grows exponentially with the increase of the desired transmission data. SOS has reduced this term as the results show. The transmission energy consumption results measured by joule and acquired by M3WSN directly confirm the SOS role in saving the WMSN

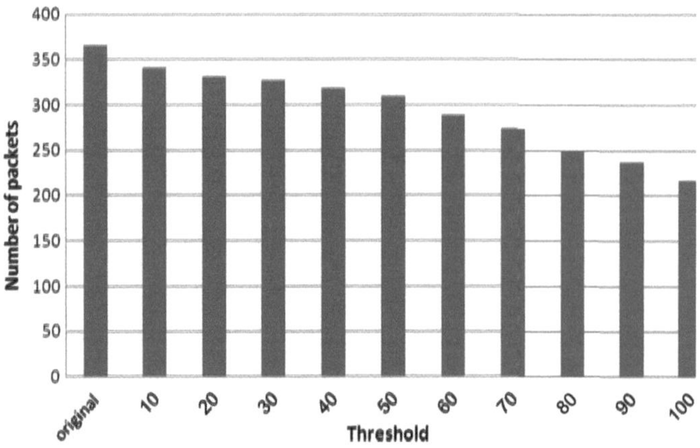

Fig. 9 Number of packets

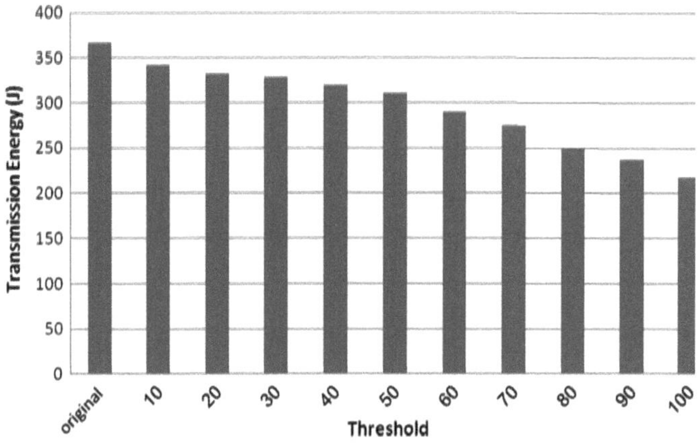

Fig. 10 Transmission energy

life. The attained energy reduction may reach 40.68%, Fig. 10. It is worth mentioning that the presented results of transmission energy already include the consumed energy for mutual control messaging.

4.4 Battery Remaining Energy

To achieve video summarization task, SOS needs only a very small additional energy consumption from the wireless multimedia sensor node. Indeed, since SOS is very simple, it consumes a fixed low amount of energy. This amount relates to a number of factors which are: frame acquisition duration, frame acquisition rate, region of

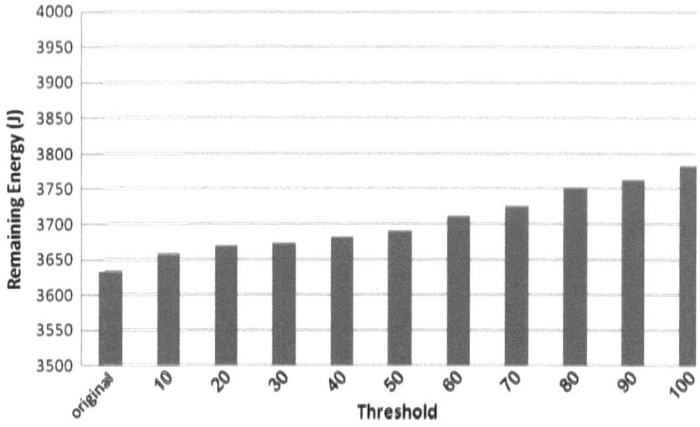

Fig. 11 Remaining energy of node battery

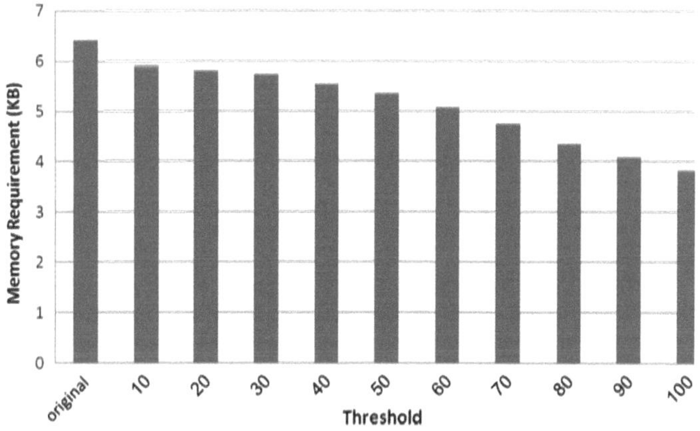

Fig. 12 Memory requirement

interest size, maximum difference calculation and the threshold value. Figure 11 shows the remaining energy in the node battery at the end of simulation. The results assure that SOS saves the node battery life. For example, SOS with threshold of 100, saves 149.324 Joules compared to no video summarization original case.

4.5 *Memory Requirements*

WMSN nodes have limited storage space. Memory requirement for storing the summarized video is lower compared to the original one as the results show in Fig. 12. SOS with threshold of 100, for example, can achieve comparable memory saving of 40.3%.

5 Conclusions

In this paper we have proposed a new Simple On-line Single-view video summarization mechanism for M2M WMSN video applications. SOS exploits the CMOS sensor structure of the sensor nodes to reduce the summarization requirement, and to save the quality of experience of the end user. We employ M3WSN framework for the simulation and evaluation of video summarization. SOS saves the limited resources of sensor nodes and remarkably improves the whole WMSN life time. Developing video summarization mechanisms that exploit the distributed field of views of multiple sensor nodes in WMSN is a promising research area.

References

1. Saveliev A, Malov D, Tamashakin M, Budkov V (2017) Service and multimedia data transmission in IoT networks using hybrid communication devices. In: 12 th International Scientific-Technical Conference on Electromechanics and Robotics "Zavalishin's Readings", St. Petersburg (Russia), vol 113, pp 1–6
2. Tannoury A, Darazi R, Guyeux C, Makhoul A (2017) Efficient and accurate monitoring of the depth information in a Wireless Multimedia Sensor Network based surveillance, arXiv:1706.08088 [cs.CV]
3. Jbeily T, Alkubeily M, Hatem I (2015) An Efficient adaptation of edge feature-based video processing algorithm for wireless multimedia sensor networks. Int J Comput Sci Trends Technol (IJCST) 3(3):156–166
4. Prabhu TN, Ranjeeth KC, Mohankumar B (2014) Energy-efficient and secured data gathering in wireless multimedia sensor networks. Int J Innov Res Comput Commun Eng 2(2):3073–3079
5. Jbeily T, Alkubeily M, Hatem I (2015) A new symmetric-object oriented approach for motion estimation in wireless multimedia sensor networks. Int J Sci Res (IJSR) 4(11):1329–1337
6. Almeida J, Leite NJ, Torres R da S (2013) Online video summarization on compressed domain. J Vis Commun Image Represent 24(6):729–738
7. Khan YS, Pawar S (2015) Video summarization: survey on event detection and summarization in soccer videos. Int J Adv Comput Sci Appl (IJACSA)s 6(11), 256–259
8. Sebastian T, Puthiyidam JJ (2015) A survey on video summarization techniques. Int J Comput Appl (0975–8887) 132(13):31–33, 2015
9. Sujatha C, Mudenagudi U (2011) A study on key frame extraction methods for video summary. Proc Int Conf Comput Intell Commun Netw, 73–77
10. Tian Z, Xue J, Lan X, Li C, Zheng N (2011) Key object-based static video summarization. Proc ACM Int Conf Multimedia, 1301–1304
11. Ou SH, Lee CH, Somayazulu VS, Chen YK, Chien SY (2015) On-line multi-view video summarization for wireless video sensor network. IEEE J Sel Topics Signal Proc 9(1): 195–179
12. Nathans J, Thomas D, Hogness DS (1986) Molecular genetics of human color vision: the genes encoding blue, green, and red pigments. Sci New Ser 232(4747):193–202
13. Ang LM, Seng KP, Chew LW, Yeong LS, Chia WC (2013) Wireless multimedia sensor networks on reconfigurable hardware: information reduction techniques. Springer Science & Business Media
14. Rosrio D, Zhao Z, Silva C, Cerqueirae, Braun T (2013) An OMNeT++ framework to evaluate video transmission in mobile wireless multimedia sensor networks. In: 6th International Conference on Simulation Tools and Techniques, pp 277–284

Design and Development of a Pipeline Inspection Robot

M. Roussialian, H. Al Zanbarakji, A. Khawand, A. Rahal and M. Owayjan

Abstract Since pipelines have become an important asset to the economic development of almost any country, maintaining and conserving them is a must. Some of the methods used to maintain them are as simple as inserting a tube with a camera to find the defective part of the pipeline or continuous digging to assess and find the problem. However, most methods currently used involve taking risks to reach inaccessible areas in a pipeline. The proposed semi-autonomous pipe inspection robot is designed to navigate through tight areas inside a pipe, with the ability to adapt to different pipe sizes and report the state of the pipe and the nature of any problem found. This robotic system adapts easily to the cylindrical shape of a pipe and runs on DC geared motors that provide the required torque for the robot to ascend vertically if needed. In addition, this particular robot does not only work by user command but can also be configured to run specific autonomous tasks. The sensor data collected is stored on a router acting as a web server that can be monitored by the user and accessed in addition to the live-feed provided by the IP camera mounted on the robot. The robot's design and control interface provide easy and safe access to tight and distant areas of a piping system. This robot has proven to solve risky, time-consuming and expensive maintenance procedures.

Keywords Automation · Mechatronics · Machine design

M. Roussialian · M. Owayjan (✉)
Department of Mechatronics Engineering, American University of Science & Technology, Beirut, Lebanon
e-mail: mowayjan@aust.edu.lb

H. Al Zanbarakji · A. Khawand · A. Rahal
Department of Computer and Communications Engineering, American University of Science & Technology, Beirut, Lebanon

R. Rizk and M. Awad (eds.), *Mechanism, Machine, Robotics and Mechatronics Sciences*, Mechanisms and Machine Science 58,
https://doi.org/10.1007/978-3-319-89911-4_4

Table 1 Hazardous liquid pipeline accident summary by case (1/1/2002-12/31/2003) [2]

Reported cause	Number of accidents	Property damages
Excavation	40	$8,987,722
Natural forces	13	$2,646,447
Weld failure	45	$30,681,741
Corrosion	69	$17,775,629

1 Introduction

A very important base of economics and trade involves having a pipelining system for transportation of goods and various materials from one destination to another, for example, Natural gas, water, etc. One of the most important aspects of maintaining such delicate transportation tools is the ongoing and continuous inspection of piping systems that need to be treated carefully, otherwise, the company involved would procure great losses in case of pipeline damage. As a solution for this sensitive issue, specific inspection systems should be deployed to contribute in fixing any possible problem that might occur. The system that is designed; uniquely stands out, since it costs significantly less than any other similar system found, a topic that is discussed later on in this report. Moreover, the robot has potential to replace basic pipe inspection tools that are limited in size, length, and capabilities.

The design of such a flexible and compact robot has not been accomplished previously. Most natural gas or water transportation companies; use bigger, more expensive and complicated versions of inspection robots which are usually limited to operate in a single pipe size. Even though various technological advancements are taking place; some noticeable issues are occurring regarding pipelines and construction sites. While digging a site, workers encounter an unexpected pipeline and damage it; without their previous knowledge of its existence [1]. This certain issue, along with others such as maintenance and natural forces; raise important questions and show a great need for certain technologies that are able to help with pinpointing a pipe's location, as well as ensuring the safety of workers. The table below, Table 1, shows the grave number of accidents and property damages resulting from a possible error made by a worker.

The main aim of the machine is to replace manned inspection of pipes, as well as increasing safety against different natural disasters. This solution also provides distant monitoring which enables the application different problem-solving techniques. In addition, this particular solution is unique in its mechanism and locomotion as it is able to move vertically up and down through pipelines. Some of the inspection robots mentioned like the PIGs offer maintenance and frequent safety check for company designated pipes. Meanwhile, the benefit offered by this solution is not only the access to a variable pipe size ranges but also the apparent reduced cost of the project [3].

Fig. 1 Robot inspecting a pipe [4]

In order to complete such an innovation, the following order of procedures is undertaken.

1. Drafting and analyzing various designs, and asking for professional opinion.
2. Part manufacturing and electronics.
3. Machine assembly.
4. Programming the controller.
5. Testing and validation.

A lot of alternative products are available in the market; each having its unique features. Some of the products worth mentioning are, The PIG robot; a relatively big sized pipe inspection robot mainly used for gas pipes. It depends on gas flow inside the pipe in order to move along the pipeline system. The robot is capable of emitting sonar signals along the pipe, in order to detect corrosion like abnormalities. This robot is quite expensive and not compact. Another inspection product is the camera and wire traditional utility; it is inserted into a pipe where the user monitors the feed provided by the camera. This traditional tool is inefficient due to having a wired connection with limited length and flexibility. The final product is the Versatrax robot, which is mainly used for climbing vertically. This robot is marginally heavier than others, dependent on a hydraulic cable for expansion, and unable to cross corners due to its design.

Different pipe sizes depend on its type and manufacturing country. Natural gas pipes are manufactured between 10 and 300 ft. depending on the company, the diameter ranges between 0.622 and 7.981 inches. While water pipes are manufactured between 1 and 30 m in length and diameter ranging between 12 to 150 mm. Regarding oil pipes, they are of length 371 km since they are mainly used to link between different countries with a diameter of 30 inches. The figure below shows a typical model of an inspection robot, Fig. 1.

Robots aim at taking over risky procedures traditionally done by humans. In the following paper [5] the author discusses the kinematic analysis and locomotion strategy of a pipe inspection robot concept for operation in active pipelines. The wheeled robots which have a simple design and advantage in power consumption can travel horizontally or slightly angled slopes; however, they are limited when it comes to high incline angles such as the case of traversing vertically upwards. Moreover, the fluids running inside pipes can become obstacles for such robots and might even damage them in certain cases. In order to overcome these obstacles, few wheeled pipe inspection robots have attempted to use an active method of attracting the wheels to the pipe wall using magnetism, or by applying forces on the walls of the pipes using springs, actuators, or a combination of both.

In spite of their mechanical simplicity, the efficiency of wheeled robots whilst climbing is not optimal, as the force used to push the wheels against the pipe wall acts as an opposing force against the actuators driving the wheels. The inchworm robots are also easy to control and they have the capability of freely navigating a pipe either by using vibration or by pressing on the sides of the pipe in different ways by either extracting or retracting the limbs. This type of robot is usually most effective in climbing but usually slower in movement. Another type of robot is the snake robot which usually has more complexity in terms of control and movement having greater degrees of freedom. The movement results from a traveling locomotion wave coming from multiple actuated joints that are connected together. It can be used inside pipes that have a 90-degree incline; as well as inside a network consisting of straight and curved sections to assess its ability to adapt to changing pipe trajectories. The robot uses an inchworm locomotion method to move inside the pipe, expanding and contracting the upper and lower rings in order to attach to and detach from the pipe wall, while using the remaining six actuators to translate and rotate the rings in six degrees of freedom to the required position for the next movement. This sensor-less system measures the amount of current being drained by the motors to calculate how much extension force is still needed to have a good friction between the wheels and the side of the pipe. Whenever the approximate force or current drain is achieved, the extension stops.

In [6] the author is taking another perspective on analyzing and controlling a pipe inspection robot that is seen in Fig. 2. It consists of a fore and rear leg system and a body. These systems are designed by using a three worm gear system arranged at angle 120 degrees to operate at different diameters inside a pipe. The springs are attached to the body and the legs to operate in pipes of diameter range between 140 to 200 mm. Many pipelines are used in chemical industries for carrying petrol, diesel or oil but throughout the years these pipes get damaged and have defects. It is a difficult task for humans to search for the defects without technological assistance. Thus, by using inspection robots, it is possible to find answers in a short period of time and at a low cost.

Regarding wireless communication, several tests were conducted on certain pipes; some with a 30 cm diameter; after calculations, the resulting internal pipe frequency was 2.4–5.2 GHz for a distance of 100 m. Therefore, a rotating probe using piezo elements and a touch sensor was developed to get the inside information of the pipes. Tests were also conducted for different types of pipes, such as steel pipes

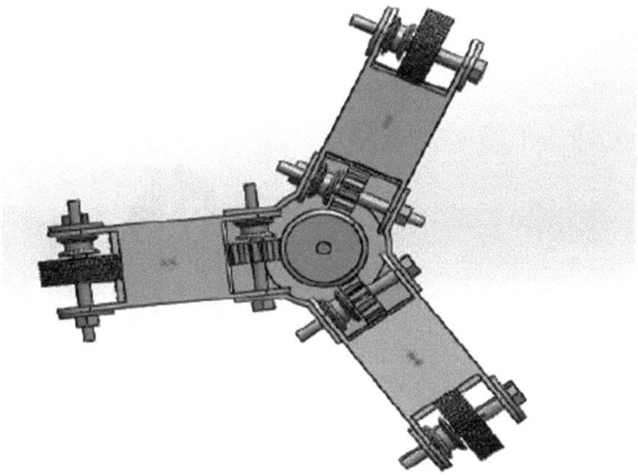

Fig. 2 Fore and rear leg system [6]

which resulted in a loss in transmission. Whereas in case of ceramic pipes, it was proven that they have good properties for wireless transmission.

2 System Design

2.1 Robot Features and Constraints

As mentioned previously, the robot has several unique features that aim at solving obstacles faced by other inspection robots in the field. The followings are its main features:

- Capability of going through horizontal and vertical pipelines.
- Wireless monitoring and control.
- Measurement of traveled distance.
- Application on different range of pipe sizes.

In spite of its many features, the robot has certain limitations such as:

- Unsuitable for water pipes.
- Unsuitable for oil pipes due to low friction.
- Limited range of pipe diameters (ranging from 25 to 50 cm).
- Limited wireless communication (maximum of 120 meters).

Fig. 3 3D Rendered model

2.2 Mechanical Design

Autodesk Inventor Professional 2015 software is used as a tool for the mechanical design of the robot. Inventor Professional 3D CAD software offers an easy-to-use set of tools for a three-dimensional mechanical design, documentation, stress analysis, and even simulation. The first step in the mechanical design is to make 3D models of available parts to be used in the project. The following step is to design the new parts and start assembly. Assembling the parts together helps determine their relations and taking projections as references. The figure below shows the 3d rendered model of the robot, Fig. 3.

2.3 Electrical Design

The electronic platforms used in the robot are an Arduino Board which is an open-source computing platform grounded on a microcontroller board, Fig. 4, a Motor Shield used to drive two 12 V geared DC motors in two directions using a special chip that can handle a current up to 2 A on each channel, Multi Turn Variable Resistors, Fig. 5, acting as shaft encoders to provide the number of turns and calculate the speed of locomotion and distance traveled. An ultrasonic sensor platform is also mounted on the front face of the device helping in obstacle avoidance. Finally, a router is placed internally with the aim of providing wireless communication.

As for the power, 11.1 V rechargeable Lithium Ion battery having a capacity of 4200 mah is used in addition to DC down converters to supply the specific voltage for each component, such as supplying constant 12 V to the motors and 5 V to the microcontroller.

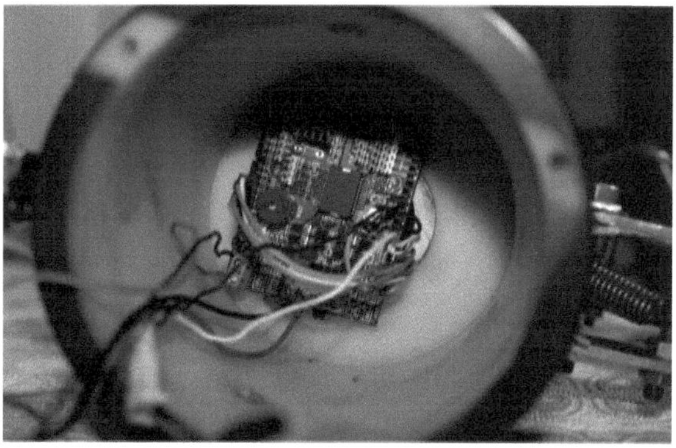

Fig. 4 Controller mounted internally

Fig. 5 Multi turn variable resistor

2.4 Software

As for the software and control, after inserting the robot into the pipe, it is controlled using the laptop; the robot either moves forward or backward depending on the command given. During locomotion, the robot measures the distance traveled using the multi turn resistors mounted on the upper arm. The calculation is done by counting the number of turns and multiplying them by $2\pi R$ where R is the radius of the wheel (equal to 6.5 cm). In addition, the robot sends a live camera feedback to a tablet that has the Show Mo application installed and notifies the user on the web server in case of any obstacle in the way of the robot by using the ultrasonic module located above the camera.

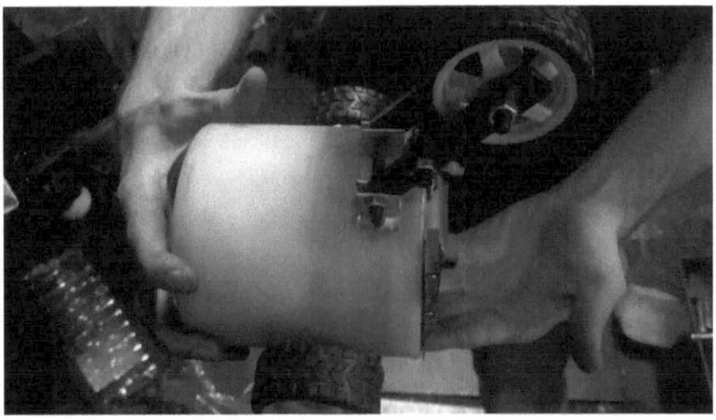

Fig. 6 Half body

Fig. 7 Motor assembly

3 Implementation and Testing

In order to start the manufacturing process, 2D documents are extracted from the 3D model and taken to a manufacturing company. The body is made up of two identical figures that can be assembled together using joints as seen in Fig. 6 below. As for the wheels, custom steel linkages are made for assembly.

After manufacturing all the different parts and purchasing the electronic equipment, each component is assembled in place starting with the motors as seen in Fig. 7 below. After having the robot assembled, a manual test is done with the aim of checking the locomotion and calibrating the linkages accordingly. The test is done using a transparent pipe of acrylic glass material, Fig. 8.

After the initial testing, the robot is disassembled for painting and neat wiring. A battery charging board is also added to externally charge the battery. Afterwards, the robot is tested again for efficiency and control at different angles. Wireless communication in terms of wireless control, data transmission, and live camera feed is also tested, Fig. 9.

Fig. 8 Initial test

Fig. 9 Advanced test

4 Conclusions and Future Work

Based on the work done, it is concluded that this design has certain advantages over other mechanisms. The system is capable of crawling inside horizontal and vertical pipelines of diameters ranging from 25 to 50 cm, measuring the distance traveled, and providing live feedback and monitoring for the user. Such robots currently in the market are bigger in size, have more limitations, and are quite expensive. The project took around 8 months to complete. This system is open for research and improvements, where more features can be added, and most; if not all constraints can be removed. Water and liquid resistance might be a crucial upgrade which would provide the robot with the capability of submerging inside water and oil transporting pipelines. Regarding radio communication, the current range can be increased dramatically by using a SCADA networking system which stands for Supervisory Control and Data Acquisition. SCADA is a system that operates with encrypted signals over many types of communication channels and can measure multiple forms of data such as flow, pressure and temperature using Remote Terminal Units. Finally, in terms of manufacturing, molds can be made for mass production using plastic injection method; this can further reduce cost and production time.

References

1. Primis.phmsa.dot.gov. Pipeline Construction|FAQs (2016). https://primis.phmsa.dot.gov/construction/faqs.htm. Accessed 18 Jan 2016
2. Corrosion-doctors.org. Pipeline failure causes (2016). http://www.corrosion-doctors.org/Pipeline/line-failures.htm. Accessed 22 Jan 2016
3. Maitre M (2015) PIGs in the Pipes|Trans Canada CorporationBlog. Blog.transcanada.com. https://blog.transcanada.com/pigs-in-the-pipes/#sthash.ckJWQ5U9.RDeunmk5.dpbs. Accessed 11 Dec 2015
4. Engineeringtoolbox.com. Natural Gas—Pipe Sizing. (2016). http://www.engineeringtoolbox.com/natural-gas-pipe-sizing-d_826.html. Accessed 28 Jan 2016
5. Bekhit A, Dehghani, A, Richardson R (2012) Kinematic analysis and locomotion strategy of a pipe inspection robot concept for operation in active pipelines. ijmem.avestia.com (2012). http://ijmem.avestia.com/2012/. Accessed 18 Dec 2015
6. Gargade A, Tambuskar D, Thokal G (2013). http://www.ijetae.com (ISSN 2250-2459, ISO 9001:2008 Certified J 3(5)). Accessed 18 Dec 2015

Developing a Sensorless Sun Tracker for PV Panels

B. Nehme, A. Fenianos and T. Akiki

Abstract Solar energy market is growing rapidly and presenting a high energy share. We can find many applications of solar energy harvesting systems like Photovoltaics, Dual Face Photovoltaics, Concentrated Photovoltaics, and Concentrated Solar Power. All these systems need tracking technology to increase their efficiency or total output. We present in this paper an innovative sensorless tracking algorithm for PV panels where we track the sun's position without using a sensor. The algorithm has been implemented on a Beckhoff PLC and a small PV module has been used to test the system.

Keywords Solar tracker · Sensorless control · Photovoltaic panels
Azimuth angle · Elevation angle

Nomenclature

LIRA	Lebanese Industrial Research Achievements Program
MPP	Maximum Power Point
MPPT	Maximum Power Point Tracker
PLC	Programmable Logic Controller
d	Julian day number
H	Solar hour angle degrees
LSTM	Local Standard Time Meridian
LT	Local time in hours
P	Local solar time

B. Nehme (✉) · A. Fenianos · T. Akiki
Holy Spirit University of Kaslik (USEK), Jounieh, Lebanon
e-mail: becharanehme@usek.edu.lb

A. Fenianos
e-mail: anthony.y.fenianos@net.usek.edu.lb

T. Akiki
e-mail: tildaakiki@usek.edu.lb

R. Rizk and M. Awad (eds.), *Mechanism, Machine, Robotics and Mechatronics Sciences*, Mechanisms and Machine Science 58,
https://doi.org/10.1007/978-3-319-89911-4_5

PV	Photovoltaic
S	Grid-shadowing factor
TC	Time correction factor
T_{GMT}	Time in Greenwich Mean Time
X	Equation of Time
α	Elevation angle of the sun in degrees
α_p	Elevation angle of the panel in degrees
β	Azimuth angle of the sun in degrees
β_p	Azimuth angle of the panel in degrees
δ	Declination angle in degrees
l	Wavelengths with energy higher than the energy gap
θ	Incident angle of the sun's rays to the panel in degrees
φ	Latitude
ω	Longitude
r(l)	Reflectance
α(l)	Coefficient of absorption
f(l)	Incident flux of the photon

1 Introduction

Nowadays, energy solutions, especially the green and the clean energy (renewable), are becoming more communal. PV (Photovoltaic) systems are a great and reliable energy harvesting solution. In order to maximize the PV systems output power, we need to tilt them to the direction of the sun. To reach this target, a tracking system is required. This is an effective solution better than getting extra solar panels to increase the output power and therefore it reduces the cost of our installation. By using a dual-axis tracker it has been projected that the income for solar panels can be improved by 40% for solar panels relatively to a fixed system.

In [1], Hao Wang et al. used a linear motor to orient the panel in the elevation angle and 4 photoresistors as a sensor. In [2], Hongyi Wang et al. used a novel sensor based on two photodiodes separated by a mental wall. The sun's angle can be determined by the two currents of the photodiodes. In [3], Djilali Chogueur et al. used a cell to measure the output power of two different positions and accordingly, they rotate the panel left or right. In some conditions, a time delay of 4 min is scheduled between two positions which may lead to wrong tracking due to change in the meteorological conditions. In [4] Hassan Fathabadi perturbed the two axis angles and compared the two MPP (Maximum Power Point) powers before and after the perturbation and decided accordingly the future change.

This paper will show an innovative tracking system that aims to keep the solar panels aligned with the rays of the Sun to maximize the output power. We start by studying the motion of the Sun; its azimuth and elevation angles. Then, we describe our algorithm. Later, we simulate our system under Matlab. At the end, we describe the electro-mechanical design of our system and its implementation.

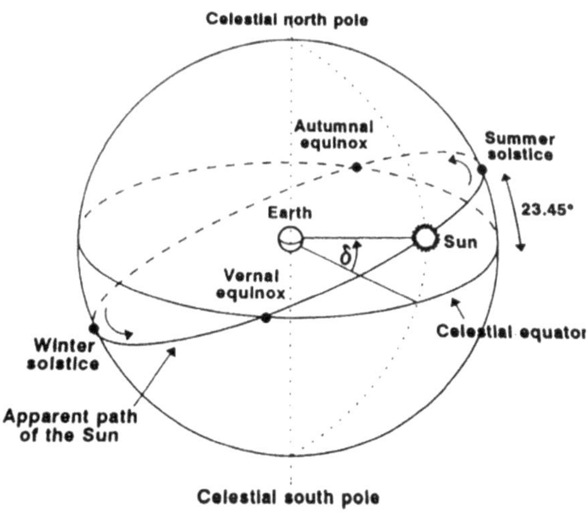

Fig. 1 Sun declination angle [6]

2 Motion of the Sun

The movement of the sun is due to the rotation of the Earth around its axis. When the sun is moving during the day and during the seasons, the angle at which the incident light hit a general point on earth changes. If we take a fixed location on earth, we can note that the sun is moving along the sky during the day. The location of the sun depends upon the position of a fixed point on the Earth, the local time, the time difference between local and Greenwich Mean Time (GMT) and the specific day of the year [5].

2.1 Declination Angle

Declination is the angle between the equator and the sun as showed in Fig. 1. The declination angle, δ, leads to a variation each period of time due to the circulation of the Earth around the sun and around its axis of rotation. If the Earth would not rotate over its axis of rotation, then the declination angle is equal to zero (0°). However, the Earth rotates 23.45° below and above the equator so the declination angle would have a range of 47°. But in two seasons, Spring and Autumn, the declination angle is equal to 0° [6].

The declination angle is given by the following formula [7]:

$$\delta = \sin^{-1}\left(\sin(23.45)\sin\left(\frac{360}{365}(d-81)\right)\right) \tag{1}$$

2.2 Elevation Angle

The elevation angle is the angle between a horizontal plane at a certain position on earth and the height of the sun. The elevation is 0° at sunrise and it is 90° when the sun is directly overhead (this case happens at the equator). It depends from the day of the year and the latitude of a specific location.

The elevation angle is given by the following formula [7]:

$$\alpha = \sin^{-1}\left(\cos(\delta)\cos(\varphi)\cos(H) + \sin(\delta)\sin(\varphi)\right) \tag{2}$$

2.3 Azimuth Angle

The path of the sun during the day is from the east at sunrise to the west at sunset. We define the azimuth angle as the deviation from the north axis. The azimuth is 90° at sunrise and 270° at sunset (for a location at the northern hemisphere).

The azimuth angle is given by the following formula [7]:

$$\beta = 180 - \cos^{-1}(\beta) \quad \text{if} \ \sin(\beta) < 0 \tag{3}$$

$$\beta = 180 + \cos^{-1}(\beta) \quad \text{if} \ \sin(\beta) > 0 \tag{4}$$

where,

$$\cos(\beta) = (\sin(\varphi)\sin(\alpha) - \sin(\delta)) \,/\, \cos(\varphi)\cos(\alpha) \tag{5}$$

$$\sin(\beta) = \cos(\beta)\sin(H)/\cos(\alpha) \tag{6}$$

We associate the following formulas to calculate the solar hour angle H [8]:
Local Standard Time Meridian

$$LSTM = 15 \times (LT - T_{GMT}) \tag{7}$$

Equation of Time

$$X = -7.53\cos(A) - 1.5\sin(A) + 9.87\sin(2A) \tag{8}$$

$$A = (d - 80)360/365 \tag{9}$$

Time Correction factor

$$TC = 4(\omega - LTSM) + X \tag{10}$$

Local Solar Time

$$P = TC/60 + LT \tag{11}$$

Solar hour angle

$$H = 15 \times (P - 12) \tag{12}$$

3 Sun Tracking Algorithm

From the equations above we can see that we can track the position of the sun knowing the time and the location of the PV panel. However, the latter means inaccuracy in cases of position change, bad operational time logging, or bad real-time clock. While using sensors in PV tracking systems, higher accuracy is obtained. However, sensors may present their own problems like crashes and price. Additional circuitry and hardware are also required when using sensors.

We propose in this paper a sensorless, offline, and robust sun tracker algorithm for PV panels. Only the controller, current and voltage probes of the MPPT (Maximum Power Point Tracker) are used.

3.1 Electron-Hole Generation Rate

For an incident light at the front surface, the electron-hole generation rate of a solar cell is given by the following equation [9]:

$$G(x) = (1 - s) \int_{\lambda} (1 - r(\lambda))\, f(\lambda)\alpha(\lambda)e^{-\alpha(x+W_N)}\, d\lambda \tag{13}$$

We can see that the electron-hole generation or the photo-generated current is proportional to the incident flux of photons f(l). The more we tilt the panel to the sun's position, the more photo-generated current we can get. From this assumption, the photo-generated current will give us an indication if the panel is being tilted correctly towards the sun position. We can build our algorithm by perturbing the panel's position and studying the change in the current.

3.2 Algorithm Development

For each axis alone, we tilt its angle for one degree, and we compare the new value of the photo-generated current. If the current has increased, we are approaching towards the sun's position; if the current has decreased, we are distancing from

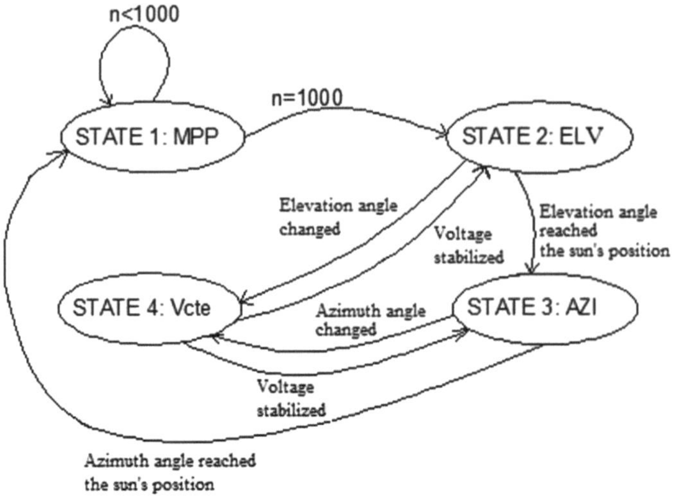

Fig. 2 Petry net of MPP and sun's position tracker

the sun's position. The photo-generated current cannot be measured directly unless short-circuit conditions are established which are not feasible because the panel is assumed to be controlled at MPP.

The panel controlled at MPP may lead in certain conditions to a decrease in current even if the sun's position has been approached. To solve this problem and to avoid going to short-circuit conditions, we fix the voltage at its current value when seeking the sun's position.

The MPPT and sun's position will be combined and the global algorithm is shown in Fig. 2. The controller operates at MPP (STATE 1). After a certain delay (defined here by the counter n = 1000), the controller will save the values of the operating voltage and current, and starts seeking the elevation angle (STATE 2). After each perturbation, the voltage is reset to its MPP value (STATE 4) and the current is compared to its previous value. Once the elevation angle reaches the sun's position, the same operation is performed for azimuth angle (STATE 3). At the end, the controller goes back to seek the MPP (STATE 1).

The MPPT algorithm used is the P&O (Perturb and Observe). We have used this standard MPPT since it is used as reference to compare MPPT algorithms and we do not intend to focus on new MPPT algorithms in this paper.

3.3 Elevation and Azimuth Algorithms

The flow chart of the elevation angle is shown in Fig. 3. The algorithm shows how the elevation angle of the PV panel (α_p) will track the sun. First, we measure the current at an unknown elevation angle. Then the elevation angle will be incremented

Fig. 3 Tracking algorithm for the elevation angle

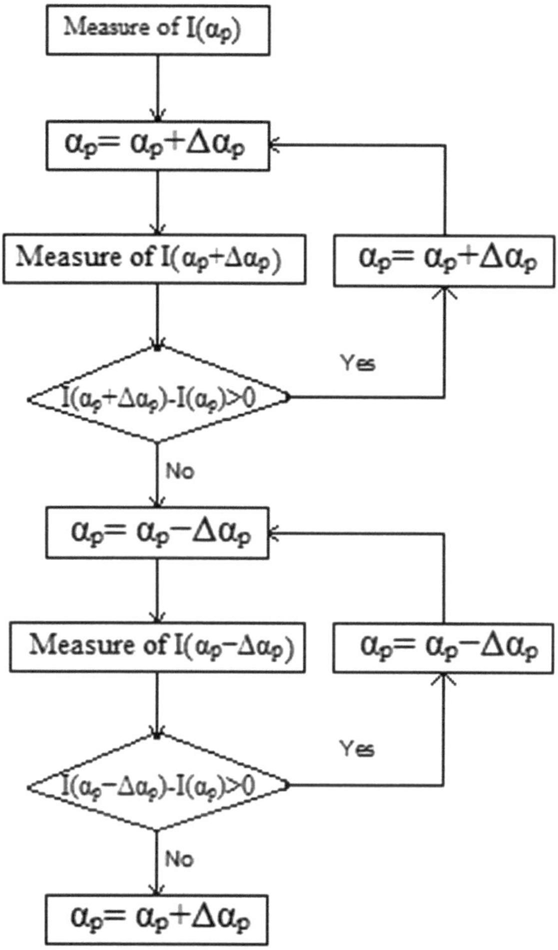

and the new current will be measured. If the difference between the output currents is higher than 0, it means that we are approaching towards the sun's position, the system continue increasing the elevation angle. Otherwise, the elevation angle will decrease to reach the maximum output current of the system. Finally, the elevation angle is increased because the last loop has been exited by a decrease in current. The convergence of this Perturb and Observe algorithm has been proved in [10].

When the elevation angle changes and we gain power, it means that the system is rotating in the correct way to be aligned with sun rays. The same algorithm is used for the azimuth angle (β_p).

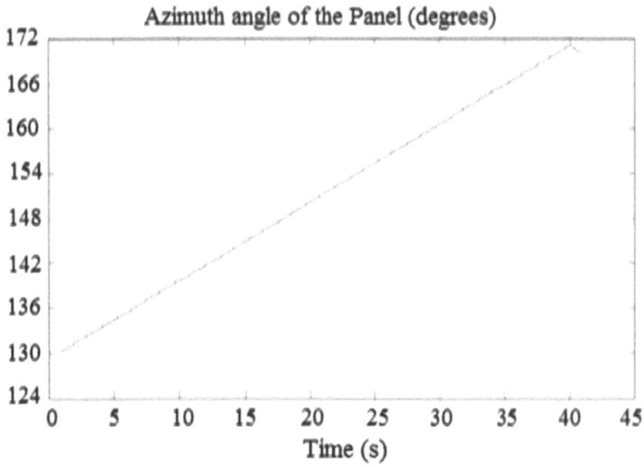

Fig. 4 Azimuth angle of the panel under the tracking algorithm

4 Matlab Simulation

We implemented the above equations stated in paragraph 2 and the algorithms developed in paragraph 3 into Matlab/Simulink. And we implemented the physical model of the photovoltaic panel. In addition, we implemented the following formula that gives the solar incident rays angle as a function of the sun position and panel position [7]:

$$\theta = \cos^{-1}\left(\cos\left(\alpha_p\right)\cos(Z) + \sin\left(\alpha_p\right)\sin(Z)\cos\left(\beta - \beta_p\right)\right) \quad (14)$$

$$Z = 90 - \alpha \quad (15)$$

We choose a location in Lebanon where the latitude is $\varphi = 35.53°$, the longitude is $\omega = 33.9°$ and the local time is 11:38 AM of the first of july (d = 182). The initial position of the panel is set to $\alpha_p = 45°$ and $\beta_p = 131°$. The tracking angle of the azimuth axis and the correlated current are shown in Figs. 4 and 5. We can see how the azimuth angle of the panel reached the azimuth angle of the sun $\left(\beta_p = 170°\right)$. In addition, the last step corresponds to the repositioning of the panel after the detection of the decrease in current.

We can see also the cosine shape of the current curve. In fact, it is related to the loss factor cos(θ), where θ is the incident angle of the sun's rays to the panel.

5 PLC Implementation

The implementation of this project has been executed on a Beckhoff PLC controller. A beckhoff PLC was chosen as a controller that can handle high temperatures as PV panels will face these conditions. A small PV panel composed of 4 cells has been mounted on an Aluminum chassis with two degrees of freedom. One horizontal axis

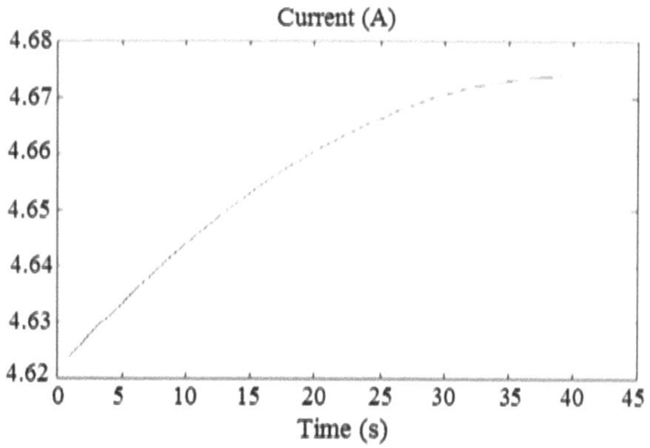

Fig. 5 Current of the panel under the tracking algorithm

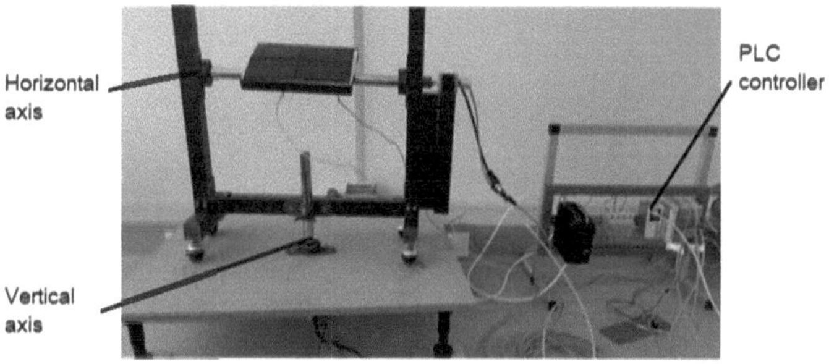

Fig. 6 Hardware and software implementation of the project

is rotated by a stepper motor and another vertical axis is rotated by another stepper motor mounted under the table as shown on Fig. 6. In addition, a buck-boost DC/DC converter has been developed for MPP and constant voltage control.

The PLC system contains the following I/O modules necessary for the communication with the hardware:

- The EL2502 output terminal for generating a PWM signal
- The EL5151 EtherCAT Terminal is an interface for the direct connection of incremental encoders built in the stepper motors for getting its position
- The EL7031 which is a stepper motor terminal for the horizontal axis
- The EL7041 which is a stepper motor terminal for the vertical axis
- The EL3681 EtherCAT Terminal enables measurement of currents and voltages of the panel

The algorithm has been implemented on the PLC. The step angle used is 1 degree. And a projector has been used to emulate the sun and its position. The projector has been positioned at an arbitrary position after the execution of the software, the panel is positioned at the best tilt angle to harvest the maximum power.

6 Discussion

The algorithm developed in this paper present a sensorless method to track the position of the sun for PV panels. The same hardware used for MPPT algorithm is used where only a change in the code is applied as shown in Fig. 2. The importance of this tracker is beyond the simplification of the traditional sensor used in such kind of systems. In fact, it addresses special types of PV panels: dual face PV panels. These panels can harvest energy from both sides, front and back. A traditional tracking system will look for positioning the panel to get a zero incident angle. However our tracker will seek the maximum power that can be generated from such kind of systems, where the optimum tilt angle does not necessarily coincide with the zero incident angle.

7 Conclusion

In this paper, we showed a new tracking algorithm for PV panels. The algorithm is sensorless and uses only existing hardware of the MPPT. It is based on the fact that higher current will be generated when the panel is tilted to the position of the sun. The algorithm has been implemented in Matlab/Simulink and on real hardware controlled via a Beckhoff PLC. In both cases, convergence has been recorded.

As future works, we intend to validate our approach, where position sensors can be implemented to the developed hardware to check that the final position of the panel converges to the sun's position. We also intend to find the optimal delay between two position adjustments (defined previously by $n = 1000$). This requires an optimization process between the energy consumption of motors and the irradiance level of the sun.

Acknowledgements The authors would like to thank the **HCR (Higher Center for Research)** of the Holy Spirit University of Kaslik for financing this project. In addition, the authors would like to thank the LIRA research grant for participating in funding this project and ITEC company for providing the Beckhoff PLC as well as Eng. Carlos Bou Gerges for his valuable help.

References

1. Li Z, Luo J, Xie S, Li H, Wang H (2013) Design and implementation of a dual-axis sun-tracking system based on microcontroller. In 2013 International conference on information technology and applications
2. Fan Y, Lu Z, Luo T, Wang H (2014) A self-powered sun tracking system based on a novel lignt direction sensor. In: 2014 12th IEEE international conference on solid-state and integrated circuit technology (ICSICT)
3. Bentouba S, Merouane A, Chogueur D (2015) Smart sun tracking system. In: 2015 3rd International renewable and sustainable energy conference (IRSEC)

4. Fathabadi H (2017) Novel online sensorless dual-axis sun tracker. IEEE/ASME Trans Mechatron 22(1):321–328
5. Brooks W, James D (2012) Photovoltaic (PV) installer resource guide. In: NABCEP
6. Markvart T (1994) Solar electricity. Wiley
7. Uysal S, Jazayeri M, Jazayeri K (2013) MATLAB/simulink based simulation of solar incidence angle and the sun's position in the sky with respect to observation points on the Earth. In: International conference on renewable energy research and applications
8. The sun's position http://pveducation.org/pvcdrom/2-properties-sunlight/suns-position. Accessed 10 Feb 2017
9. Hegedus S, Luque A (2011) Handbook of photovoltaic science and engineering. Wiley
10. M'Sirdi NK, Nehme B, Abarkan M, Rabbi A (2014) The best MPPT algorithms by VSAS approach for renewable energy sources (RES). In: 2014 3rd International symposium on environmental friendly energies and applications (EFEA)

Part III
Biomedical Engineering

Modelling the Deformation of Biologically Inspired Flexible Structures for Needle Steering

T. Watts, R. Secoli and F. Rodriguez y Baena

Abstract Recent technical advances in minimally invasive surgery have been enabled by the development of new medical instruments and technologies. To date, the vast majority of mechanisms used within a clinical context are rigid, contrasting with the compliant nature of biological tissues. The field of robotics has seen an increased interest in flexible and compliant systems, and in this paper we investigate the behaviour of deformable multi-segment structures, which take their inspiration from the ovipositor design of parasitic wood wasps. These configurable structures have been shown to steer through highly compliant substrates, potentially enabling percutaneous access to the most delicate of tissues, such as the brain. The model presented here sheds light on how the deformation of the unique structure is related to its shape, and allows comparison between different potential designs. A finite element study is used to evaluate the proposed model, which is shown to provide a good fit (root-mean-square deviation 0.2636 mm for 4-segment case). The results show that both 3-segment and 4-segment designs are able to achieve deformation in all directions, however the magnitude of deformation is more consistent in the 4-segment case.

Keywords Mechanical modelling · Deformable structures · Biomimetics
Medical devices · Needle steering

T. Watts · R. Secoli · F. Rodriguez y Baena (✉)
Imperial College London, London, UK
e-mail: f.rodriguez@imperial.ac.uk

T. Watts
e-mail: t.watts14@imperial.ac.uk

R. Secoli
e-mail: r.secoli@imperial.ac.uk

R. Rizk and M. Awad (eds.), *Mechanism, Machine, Robotics and Mechatronics Sciences*, Mechanisms and Machine Science 58,
https://doi.org/10.1007/978-3-319-89911-4_6

Fig. 1 Woodwasp, *Sirex noctilio*, boring into wood with its ovipositor to lay eggs. *Credit* Natural Sciences Image Library of New Zealand

1 Introduction

1.1 Clinical Motivation

With the advent of advanced medical devices, traditional open surgical procedures are increasingly being replaced with minimally invasive (MI) alternatives. This shift in surgical approach has been brought about by technical advances in minimally invasive techniques, achieving reduced patient trauma, shorter recovery time, lower chance of infection and reduced scarring [1]. A common MI method is percutaneous intervention, used for clinical access in procedures such as drug delivery, biopsy and brachytherapy, to name but a few. Steerable needles are a promising technology, developed to provide greater targeting accuracy and access to clinical targets which are unreachable via rigid instruments. Crucial to steering these needles, is the ability to predict their behaviour and interaction with soft-tissue.

1.2 Biological Inspiration

Engineers have often looked to emulate or draw inspiration from biological systems, which via the process of evolution have developed ingenious solutions to the problems of sensing, locomotion, manipulation, etc. In the quest to find a safe place for their offspring, certain parasitic wasps have evolved to possess a long slender channel, called an ovipositor, for depositing their larvae deep within a host. The Sirex noctilio wasp, pictured in Fig. 1, uses its multi-segment ovipositor to penetrate the bark and bore through the wood of conifers to lay its eggs [2] (Fig. 2).

Parittotokkaporn et al. [3] were the first to take inspiration from the reciprocating motion of the ovipositor, whereby different segments are inserted alternatively, and developed a needle which was able to achieve tissue traversal with a lower insertion

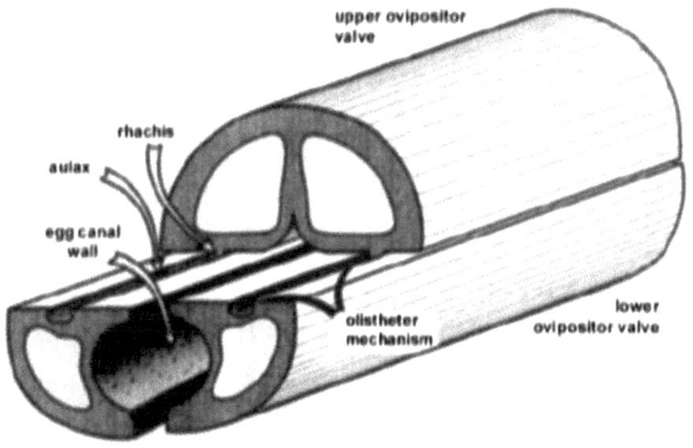

Fig. 2 Structure of the ovipositor, showing the olistheter mechanism which constrains relative motion to the axial direction [20]

force. Frasson et al. [4] subsequently demonstrated that a 2-segment needle design could perform 2D steering through a soft substrate with a controllable curvature, and later Secoli et al. demonstrated that 3D steering was achievable with a 4-segment needle. A nonlinear adaptive control strategy was implemented, based on a linear model of the needle, and shown to guide the needle along 3D trajectories [5]. With the increase in number of segments, and complexity of the design, the steering behaviour was no longer well described by the linear model, which motivates the development of a mechanics-based model for the deformation of these unique structures.

1.3 Needle Steering

Numerous needle designs have been proposed for steering through soft tissue, as outlined by van de Berg et al. [6]. The steering mechanisms are classified into seven different categories: base manipulation [7], duty-cycled bevel tip (with and without pre-curve) [8–11], pre-curved stylet [12], active cannula [13, 14], programmable bevel tip [4, 15, 16], tendon actuated tip [17, 18] and most recently, optically controlled needle [19].

Finite element analysis [21, 22], beam theory [23, 24] and spring foundations [25] have all been used to derive deflection models for steerable needles. Specifically, Oldfield et al. [21] considered the deflection of the programmable bevel-tip needle design, but the analysis is restricted to 2D. The deformation model presented here is the first to consider the complex geometry of the needle design and the effects on the 3D deformation.

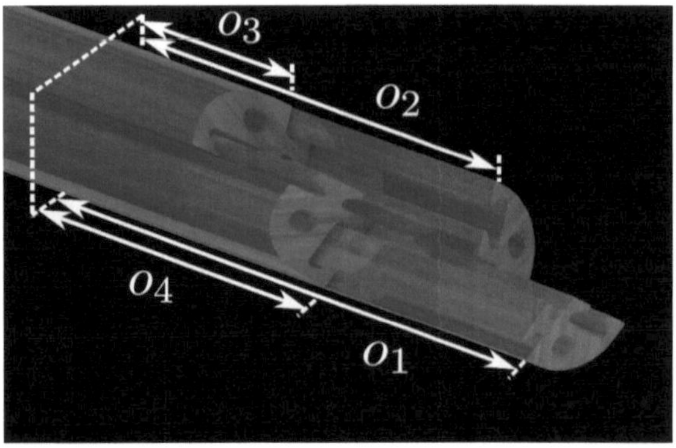

Fig. 3 Rendering of the 4-segment needle showing how the offsets o_i control the tip shape

2 Modelling Flexible Multi-segment Structures

In previous works, the shape of the needle tip has been shown to control the steering of the needle. The tip shape of an n-segment needle is determined by the relative positioning of the n segments, as shown in Fig. 3. For convenience, each segment's extension is measured relative to a reference plane that moves with the needle tip. These distances are termed the segment offsets and together form an offset vector that describes the tip configuration $o_i \in \mathbb{R}^n$. It should be noted that this representation has a single redundancy: the common-mode displacement of all segments. In the model presented here, this redundancy is given meaning, with the reference plane notionally describing the boundary between the needle tip, which controls steering, and the needle shaft, which has been shown to approximately follow the trajectory described by the tip.

2.1 System Description

A 2D representation of the beam-bending problem is shown in Fig. 4 for the 4-segment case. The tip force for each segment i, $\boldsymbol{f}_i \in \mathbb{R}^2$, is shown acting on the needle tip in the undeformed state. The needle tip is partitioned into portions of constant cross-section. Each portion j has length l_j with corresponding second moment of area $\widetilde{\mathbf{I}}_j \in \mathbb{R}^{2\times 2}$. The deflection $\boldsymbol{w}_j \in \mathbb{R}^2$ and rotation $\boldsymbol{\theta}_j \in \mathbb{R}^2$ of each portion are shown for the needle tip in the deformed state. It is also convenient to define a set J_j of segments comprising the portion (e.g. $J_3 = \{1, 2\}$). In the following sections, $\mathbf{R}_\theta \in SO(2)$ denotes a matrix that represents an anticlockwise rotation of the coordinate system by an angle θ i.e.

$$\mathbf{R}_\theta = \begin{bmatrix} \cos(\theta) & \sin(\theta) \\ -\sin(\theta) & \cos(\theta) \end{bmatrix}$$

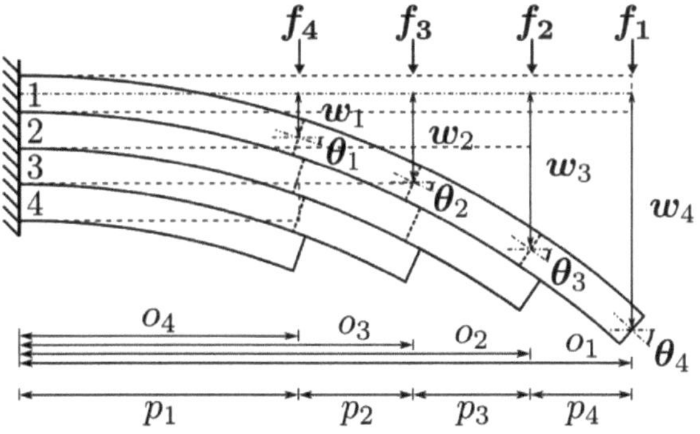

Fig. 4 2D representation of the proposed multi-segment deformation model

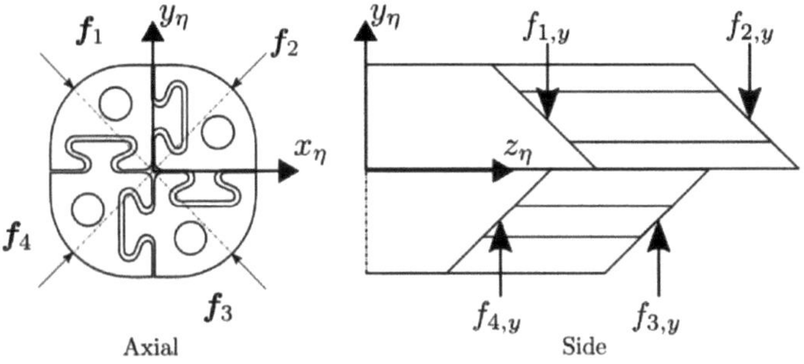

Fig. 5 Axial and side views of the transverse forces acting upon the faces at thc tips of the segments

2.2 *Tip Forces*

The point load f_i acting on each segment is assumed equivalent to the uniform pressure that would be experienced as the needle traverses soft tissue. The axial component is neglected as only transverse deflections are of interest. Figure 5 shows the cross-section of the needle, comprising the four segments, with the respective tip forces shown.

For an n-segment needle the expression for the tip force f_i in the needle tip coordinate system η is as follows:

$$f_i = \begin{pmatrix} f_{i,x} \\ f_{i,y} \end{pmatrix} = \underbrace{\mathrm{F}}_{magnitude} \underbrace{\mathbf{R}_{\frac{\pi}{n}(2i-3)} \begin{pmatrix} 0 \\ -1 \end{pmatrix}}_{direction}$$

with the force magnitude F equal for all segments. When viewed axially, the forces acting upon the needle are symmetric and sum to zero. The ability to control the direction of deformation is due to structure's configurable nature.

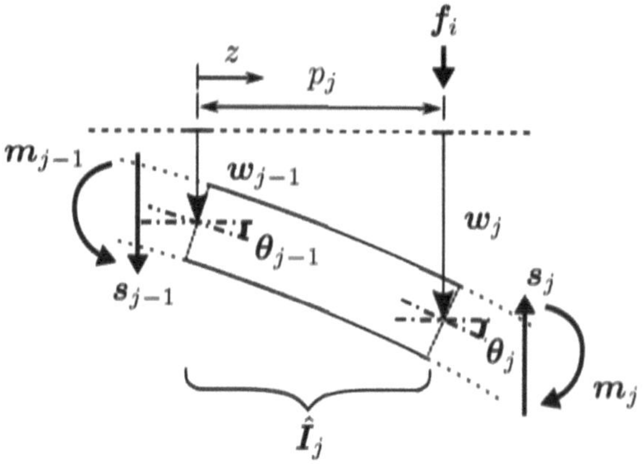

Fig. 6 Internal bending moment $\boldsymbol{m}_j$ and shear force $\boldsymbol{s}_j$ shown for portion j

2.3 Deflection Model

The second step is to determine the shear force and bending moment acting along the needle tip. Initially, they are calculated at the changes in cross section, given by the portion boundaries. In Fig. 6, a 2D representation of a portion j is shown. A virtual cut is considered at the portion boundary and the resultant shear force s_j and bending moment m_j are shown. As above, for the 3D bending s_j, $m_j \in \mathbb{R}$ are replaced by the corresponding $\boldsymbol{s}_j$, $\boldsymbol{m}_j \in \mathbb{R}^2$.

The resultant $\boldsymbol{s}_j$ and $\boldsymbol{m}_j$ are calculated as follows

$$\boldsymbol{s}_j = \begin{pmatrix} s_{j,x} \\ s_{j,y} \end{pmatrix} = \sum\nolimits_{J_j} \boldsymbol{f}_l$$

$$\boldsymbol{m}_j = \begin{pmatrix} m_{j,x} \\ m_{j,y} \end{pmatrix} = \sum\nolimits_{J_j} \mathbf{R}_{\frac{\pi}{2}} \boldsymbol{f}_l \left(o_l - \sum\nolimits_{k=1}^{j} l_k \right)$$

Note the rotation $\mathbf{R}_{\frac{\pi}{2}}$ is necessary so that a force along the x-axis corresponds to a moment about the y-axis and vice versa.

For a general cross section, the geometrical contribution to the bending stiffness is given by the second moment of area tensor

$$\mathbf{I} = \begin{bmatrix} I_{xx} & I_{xy} \\ I_{yx} & I_{yy} \end{bmatrix}$$

where I_{xx} and I_{yy} are the second moments of area about the centroidal perpendicular axes x-x and y-y respectively, and $I_{xy} = I_{yx}$ is the corresponding product moment of area.

Assuming that all segments have the same cross section geometry, it is sufficient to calculate the second moment of area tensor for a single segment $\mathbf{I}_1$. The transformed second moment of area tensor for any segment i can therefore be calculated as follows

$$\mathbf{I}_i = \mathbf{R}_{\frac{2\pi}{n}(i-1)}\mathbf{I}_1\mathbf{R}^{\mathrm{T}}_{\frac{2\pi}{n}(i-1)}$$

Assuming frictionless contact between the segments, the second moment area tensor for portion j is simply equal to the sum of the second moment of area tensors for the comprising segments, i.e.

$$\hat{\boldsymbol{I}}_j = \sum_{J_j} \mathbf{I}_i$$

The Euler-Bernoulli beam equation is as follows

$$\boldsymbol{m} = \begin{pmatrix} m_x \\ m_y \end{pmatrix} = -E\mathbf{I}\boldsymbol{\kappa} = -E\mathbf{I}\begin{pmatrix} \kappa_x \\ \kappa_y \end{pmatrix} = E\mathbf{I}\mathbf{R}_{\frac{\pi}{2}}\frac{\mathrm{d}^2\boldsymbol{w}}{\mathrm{d}z^2}$$

where s is length along the beam, $\boldsymbol{\kappa}(z)$ is curvature, $\boldsymbol{m}(z)$ is the bending moment, $\boldsymbol{w}(z)$ is deflection, E is the Young's modulus and $\mathbf{I}$ is the second moment of area tensor.

Rearranging yields

$$\frac{\mathrm{d}^2\boldsymbol{w}}{\mathrm{d}z^2} = \frac{1}{E}\mathbf{R}^T_{\frac{\pi}{2}}\mathbf{I}^{-1}\boldsymbol{m} \tag{1}$$

For $0 \le z \le l_j$ measured along a single portion j, the bending moment as a function of z is

$$\boldsymbol{m}(z) = \boldsymbol{m}_j + (l_j - z)\,\mathbf{R}_{\frac{\pi}{2}}\boldsymbol{s}_j \tag{2}$$

Substituting (1) into (2), integrating twice with respect to z and applying continuity conditions at the portion boundaries yields the expressions for the rotation and deflection of each portion.

For $j = 1, \ldots, n$

$$\boldsymbol{\theta}_j = \begin{pmatrix} \theta_{j,x} \\ \theta_{j,y} \end{pmatrix} = \left.\frac{d\boldsymbol{w}}{\mathrm{d}z}\right|_{\mathrm{z=l_j}}$$

$$= \underbrace{\boldsymbol{\theta}_{j-1}}_{previous} + \underbrace{\frac{l_j^2}{2E}\mathbf{R}^T_{\frac{\pi}{2}}\hat{\mathbf{I}}_j^{-1}\mathbf{R}_{\frac{\pi}{2}}\boldsymbol{s}_j}_{force} + \underbrace{\frac{l_j}{E}\mathbf{R}^T_{\frac{\pi}{2}}\hat{\mathbf{I}}_j^{-1}\boldsymbol{m}_j}_{moment}$$

$$\boldsymbol{w}_j = \begin{pmatrix} w_{j,x} \\ w_{j,y} \end{pmatrix} = \boldsymbol{w}|_{\mathrm{z=l_j}}$$

$$= \underbrace{\boldsymbol{w}_{j-1}}_{previous} + \underbrace{l_j \boldsymbol{\theta}_{j-1}}_{rotation} + \underbrace{\frac{l_j^3}{3E} \mathbf{R}_{\frac{\pi}{2}}^T \hat{\mathbf{I}}_j^{-1} \mathbf{R}_{\frac{\pi}{2}} \boldsymbol{s}_j}_{force} + \underbrace{\frac{l_j^2}{2E} \mathbf{R}_{\frac{\pi}{2}}^T \hat{\mathbf{I}}_j^{-1} \boldsymbol{m}_j}_{moment}$$

With fixed boundary conditions at the base:

$$\boldsymbol{\theta}_0 = \boldsymbol{w}_0 = \begin{pmatrix} 0 \\ 0 \end{pmatrix}$$

The deflection of the furthermost point of the tip is therefore given by $\boldsymbol{w}_n$, the model predicted tip deflection.

3 Finite Element Study

As a first step to evaluate the validity of the model, Finite Element Analysis (FEA) is performed on the 3-segment and 4-segment needle geometries and the observed deflections are compared with model-predicted deflections.

3.1 System Description

The segments of the structure were meshed with linear hexahedral elements of type C3D8R. A mesh convergence study confirmed stability for deformation with respect to mesh size. For the analysis, the interaction between segments is frictionless and the material is linear elastic with Young's modulus E and Poisson's ration ν. The second moment of area tensor for a segment's cross-section is $\boldsymbol{I}$. The loading at the tip of each segment is provided by a uniform pressure p, acting on the bevelled faces of area A, which are cut at an angle α. Therefore the equivalent transverse force magnitude, as defined in the model, is $F = pAcos(\alpha)$. The analysis was performed with Abaqus explicict, using the 'smooth' loading profile and the parameters given in Table 1. Deflection was measured at the tip node of the furthermost extended segment.

3.2 Results and Discussion

The resultant tip deformations for the offset configurations trialled are shown below in Table 2 for both the 3-segment and 4-segment needles.

Table 1 Parameters used in the finite element study, along with respective units

Parameter	Value (units)
E	15 MPa
ν	0.33
p	0.005 N mm^{-2}
α	30°
for n = 3	
$\boldsymbol{I}_1$	$\begin{bmatrix} 0.2582 & -0.1155 \\ -0.1155 & 0.3120 \end{bmatrix}$ mm^4
A	3.015 mm^2
F_i	0.0131 N
for n = 4	
$\boldsymbol{I}_1$	$\begin{bmatrix} 0.1082 & -0.0229 \\ -0.0229 & 0.2147 \end{bmatrix}$ mm^4
A	2.813 mm^2
F_i	0.0095 N

The tip node deflection for each configuration, $\boldsymbol{w}_n$, is shown in Figs. 7 and 8 for the 3-segment and 4-segment needles respectively, along with the deflection path as the load on each segment is increased from 0 to F_i.

The root-mean-square deviation (RMSD) between the Finite Element simulation and the model predicted values was 0.1664 mm and 0.2636 mm for the 3-segment and 4-segment cases respectively. Figures 7 and 8 both show the increased deflection for offset configurations with either a single or two segments forward, agreeing with previous observation of the needle steering [26]. Additionally, asymmetry due to the needle geometry is observed in both sets of results. The tip deflection paths are "*quasi-linear*", with non-linearity becoming a greater factor as the deflection increases.

By the comparing the deflection behaviour of the 3-segment and 4-segment structures, we can make predictions about how the steering behaviour will differ. One of the immediately discernible differences is the variation in deformation magnitude with direction. The ratio of maximum to minimum deflection is 1.703 for the 3-segment case and 1.235 for the 4-segment case, suggesting that the steering behaviour will be more consistent for the 4-segment design. This is desirable for both control and trajectory planning.

Table 2 Finite element simulation results, showing axial view of needle tip deflections. An equal load is applied to each segment tip

Tip configuration (mm) $\{o_1, o_2, o_3, o_4\}$	Deflection	Tip configuration (mm) $\{o_1, o_2, o_3\}$	Deflection
A4: {20, 20, 10, 10}		A3: {20, 20, 10}	
B4: {20, 18, 10, 10}		B3: {20, 18, 10}	
C4: {20, 16, 10, 10}		C3: {20, 16, 10}	
D4: {20, 14, 10, 10}		D3: {20, 14, 10}	
E4: {20, 12, 10, 10}		E3: {20, 12, 10}	
F4: {20, 10, 10, 10}		F3: {20, 10, 10}	
G4: {20, 10, 10, 12}		G3: {20, 10, 12}	
H4: {20, 10, 10, 14}		H3: {20, 10, 14}	
I4: {20, 10, 10, 16}		I3: {20, 10, 16}	
J4: {20, 10, 10, 18}		J3: {20, 10, 18}	
K4: {20, 10, 10, 20}		K3: {20, 10, 20}	

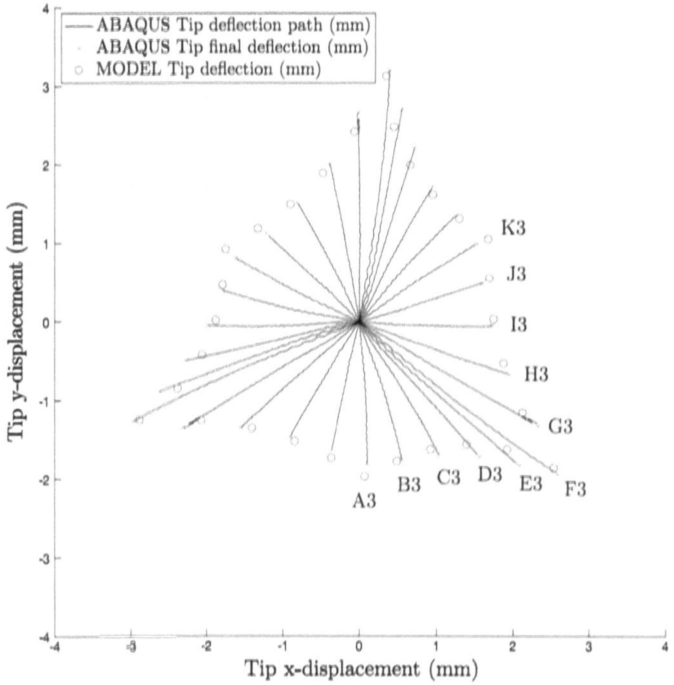

Fig. 7 Comparison between Finite Element simulation results and model predictions for the deflection of 3-segment needle tip

4 Conclusion

In this paper, we have presented a model for the deformation of flexible multi-segment structures, which have previously been shown to steer through highly compliant substrates. The model was evaluated via comparison with a set of finite element simulations, performed with 3-segment and 4-segment needle geometries, and was shown to provide a good fit, with a root-mean-square deviation of 0.2636 mm for the 4-segment case.

The results show that the relationship between the configuration of the needle tip and resultant deformation is highly nonlinear, with maximum deformation achieved when a single segment is extended. The results also show how the magnitude of deformation varies with direction, with the 4-segment needle providing more consistent deformation than the 3-segment case, suggesting this design is preferable for applications where multi-directional steering is required.

In future work, we hope to develop a full steering model for the multi-segment needle, based on the deformation model presented here. The model allows us to predict how the needle cross-section affects the system behaviour, and this will be used to inform future design iterations.

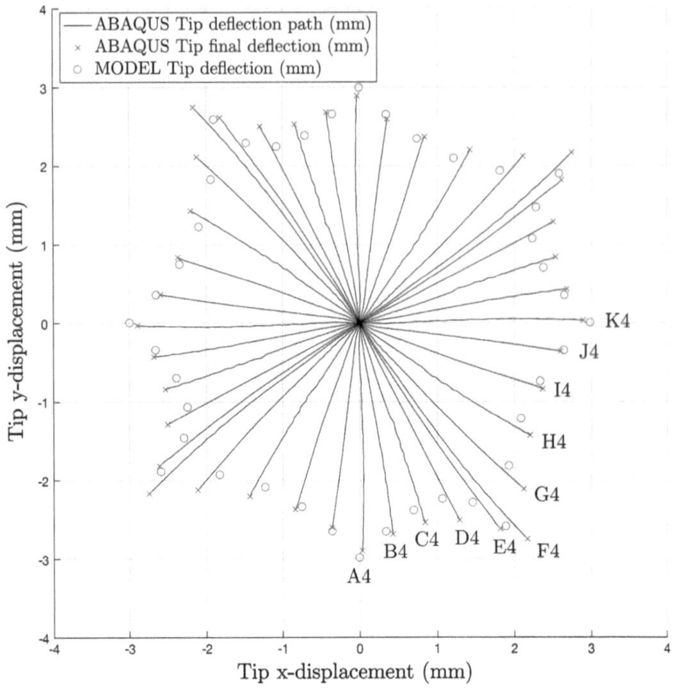

Fig. 8 Comparison between Finite Element simulation results and model predictions for the deflection of 4-segment needle tip

Acknowledgements This work was supported by the European Union's Horizon 2020 research and innovation Programme under grant agreement No 688279.

References

1. Abolhassani N, Patel R, Moallem M (2007) Needle insertion into soft tissue: A survey. Med Eng Phys 29(4):413–431
2. Vincent JFV, King MJ (1995) The mechanism of drilling by wood wasp ovipositors. Biomimetics 3:187–201
3. Parittotokkaporn T, Frasson L, Schneider A, Huq SE, Davies BL, Dagenaar P, Biesenack J, Rodriguez y Baena F (2008) Soft tissue traversal with zero net force: feasibility study of a biologically inspired design based on reciprocal motion. In: Proceedings of the IEEE international conference on robotics and biomimetics Bangkok, Thailand
4. Frasson L, Ko SY, Turner A, Parittotokkaporn T, Vincent JF, Rodriguez y Baena F (2010) STING: a soft-tissue intervention and neurosurgical guide to access deep brain lesions through curved trajectories. Proc Inst Mech Eng Part H: J Eng Med 224(6):775–788
5. Secoli R, Rodriguez y Baena F (2016) Adaptive path-following control for bio-inspired steerable needles. In: Proceedings of the 6th IEEE international conference on biomedical robotics and biomechatronics (BioRob), pp 87–93

6. van de Berg NJ, van Gerwen DJ, Dankelman J, van den Dobbelsteen JJ (2015) Design choices in needle steering—a review. IEEE/ASME Trans Mechatron 20(5):2172–2183
7. Reed KB, Majewicz A, Kallem V, Alterovitz R, Goldberg K, Cowan NJ, Okamura AM (2011) Robot-assisted needle steering. IEEE Robot Autom Mag 18(4):35–46
8. Engh JA, Minhas DS, Kondziolka D, Riviere CN (2010) Percutaneous intracerebral navigation by duty-cycled spinning of flexible bevel-tipped needles. Neurosurgery 67(4):1117–1122
9. Kallem V, Cowan NJ (2009) Image guidance of flexible tip-steerable needles. IEEE Trans Robot 25(1):191–196
10. Reed KB, Kallem V, Alterovitz R, Goldberg K, Okamura AM, Cowan NJ (2008) Integrated planning and image-guided control for planar needle steering. In: Proceedings of the IEEE conference on biomedical robotics and biomechatronics, pp 819–824
11. Webster RJ, Kim JS, Cowan NJ, Chirikjian GS, Okamura AM (2006) Nonholonomic modeling of needle steering. Int J Robot Res 25(5–6):509–525
12. Swaney PJ, Burgner J, Gilbert HB, Webster RJ (2013) A flexure-based steerable needle: High curvature with reduced tissue damage. IEEE Trans Biomed Eng 60(4):906–909
13. Dupont PE, Lock J, Itkowitz B, Butler E (2010) Design and control of concentric-tube robots. IEEE Trans Robot 26(2):209–225
14. Rucker DC, Jones BA, Webster RJ (2010) A model for concentric tube continuum robots under applied wrenches. In: Proceedings of the IEEE international conference on robotics and automation (ICRA), pp 1047–1052
15. Frasson L, Parittotokkaporn T, Davies BL, Rodriguez y Baena F (2008) Early developments of a novel smart actuator inspired by nature. In: 15th International conference on mechatronics and machine vision in practice, pp 163–168
16. Ko SY, Rodriguez y Baena F (2012) Trajectory following for a flexible probe with state/input constraints: An approach based on model predictive control. Robot Auton Syst 60(4):509–521
17. Qi P, Liu H, Seneviratne L, Althoefer K (2014) Towards kinematic modeling of a multi-DOF tendon driven robotic catheter. In: Proceedings of the 36th annual international conference of the ieee engineering in medicine and biology society (EMBC), pp 3009–3012
18. Roesthuis RJ, van de Berg NJ, van den Dobbelsteen JJ, Misra S (2015) Modeling and steering of a novel actuated-tip needle through a soft-tissue simulant using fiber bragg grating sensors. In: IEEE international conference on robotics and automation (ICRA), pp 2283–2289
19. Ryu SC, Quek ZF, Koh J, Renaud P, Black RJ, Moslehi B, Daniel BL, Cho K, Cutkosky MR (2015) Design of an optically controlled MR-Compatible active needle. IEEE Trans Robot 31(1):1–11
20. Quicke DLJ, Ralec AL, Vilhelmsen L (1999) Ovipositor structure and function in the parasitic Hymenoptera with an exploration of new hypotheses. Atti Acad Naz Ital Entomol Rc 47:197–239. http://bit.do/eeFMZ
21. Oldfield MJ, Dini D, Giordano G, Rodriguez y Baena F (2013) Detailed finite element modelling of deep needle insertions into a soft tissue phantom using a cohesive approach. Comput Methods Biomech Biomed Eng 16(5):530–543
22. Alterovitz R, Goldberg K, Okamura AM (2005) Planning for steerable bevel-tip needle insertion through 2D soft tissue with obstacles. In: IEEE international conference on robotics and automation, pp 1640–1645
23. DiMaio SP, Salcudean SE (2005) Interactive simulation of needle insertion models. IEEE Trans Biomed Eng 52(7):1167–1179
24. Khadem M, Rossa C, Usmani N, Sloboda RS, Tavakoli M (2016) A two-body rigid/flexible model of needle steering dynamics in soft tissue. IEEE/ASME Trans Mechatron 21(5):2352–2364
25. Glozman D, Shoham M (2004) Flexible needle steering and optimal trajectory planning for percutaneous therapies. In: Proceedings of the international conference on medical image computing and computer assisted intervention, pp 137–144
26. Burrows C, Secoli R, Rodriguez y Baena F (2013) Experimental characterisation of a biologically inspired 3D steering needle. In: Proceedings of the international conference on control, automation and systems, pp 1252–1257

27. Quicke DLJ, Fitton MG (1995) Ovipositor steering mechanisms in parasitic wasps of the families Gasteruptiidae and Aulacidae (hymenoptera). Proc R Soc Lond 261:99–103

On Control Design for a Lower Limb Orthosis: A Comparative Study in Different Operating Conditions

N. Roula, A. Chemori, R. Rizk and Y. Zaatar

Abstract This paper deals with the control of a lower limb orthosis. In which a model of the shank-orthosis system is given, we consider the human effort as external torque acting on the system. A comparative study, through a number of simulations in different operational scenarios, highlights the limits of a standard PID controller. On another hand, this work brings out the benefits of orienting the control strategies toward model reference approaches. Hence, a better exploitation of the nonlinear system dynamics; by deriving adaptive control strategies to enable the management of parameters uncertainty could be a relevant approach for such systems.

Keywords Orthosis · Exoskeleton · Adaptive control · PID · Sliding mode

1 Introduction

Nowadays Robotics is presented as a solution to the rehabilitation and assistance of dependent people. Wearable robots are designed to fit human anatomies. They have to ensure the best assistance to people having physical weaknesses, in their daily tasks. Also, wearable robots are used to ensure the physical and neurological rehabilitation of joints and muscles. They lead patients to recover and regain their mobility. Three case studies are considered in this paper: Passive rehabilitation concerns people who have lost control of their limbs. The second case is the assistance as needed where the wearer is involved to develop a muscular torque in the same direction as the desired trajectory. The third case is the resistive rehabilitation where the

N. Roula (✉) · Y. Zaatar
LPA, Fanar, Lebanon Lebanese University Faculty of Sciences, Beirut, Lebanon
e-mail: nadjah.roula@gmail.com

A. Chemori
LIRMM, 161 Rue Ada University of Montpellier 2, Montpellier, France

R. Rizk
Faculty of engineering, CRSI, Lebanese University, Roumieh, Lebanon

R. Rizk and M. Awad (eds.), *Mechanism, Machine, Robotics and Mechatronics Sciences*, Mechanisms and Machine Science 58,
https://doi.org/10.1007/978-3-319-89911-4_7

patient is asked to develop an effort in the opposite direction of the robots movement [1–3]. The intention of the subject can be estimated using EMG electrodes fixed at the thigh muscles [4]. Maximizing the transparency of the system to the wearer, particularly in an assistance as needed and a resistive rehabilitation contexts is one of the main problems addressed. Changes in the human-orthosis dynamics during the rehabilitation processes need to be efficiently compensated. A classical PID (proportional, integral, derivative) structure for gait assistance was designed. One of its derivatives: the proportional controller was tested. This controller amplifies the human torque measured by EMG [5, 6]. The proportional derivative controller (PD) was also developed in [7] and shoes insoles were used to measure the human. In [8], a model reference adaptive control was developed. This controller was designed in order to compensate for dynamics changes in human-orthosis system [9–11]. This technique has minimized the torque driving the system to the reference model. The control law is based on an on-line modulation of the control parameters. The writers highlighted the main advantage in using adaptive strategy which is the online parameters computation to ensure the best convergence of the trajectory. This control law is subject of a study in this paper. In [12] augmented $\mathcal{L}1$ adaptive control is developed and tested. An additional nonlinear proportional controller term is added to the torque input. The main aim of this proposed law is to mitigate the drawback due to the time lag introduced by a classical $\mathcal{L}1$ adaptive law. The paper is structured as follows. In Sect. 2, the shank-foot-exoskeleton model is presented based on previous modeling works. In Sect. 3 the three control laws: PID, Sliding Mode and a Model Reference adaptive control are presented with the stability studies related to each one. Simulation results are addressed in Sect. 4 for five scenarios. Finally, conclusions and future works are given in Sect. 5.

2 System's Modeling

EICOSI (cf. Fig. 1) is a one DOF lower limb exoskeleton. A brushless DC motor drives the wearable robots and an incremental encoder measures the knee joint angle. When embodied by the wearer and attached to his leg using straps the whole system shank-foot-exoskeleton rotates synchronously and is considered as an entity that is put in movement when one (The control Torque τ) or two torques (The control Torque τ and the Human torque τ_h) are applied to the system as external inputs. J_s,τ_{gs},A_s,B_s and K_s are respectively the shank inertia, gravity torque, solid friction coefficient, viscous friction coefficient and the stiffness coefficient. J_0,τ_{g0},A_0,B_0 and K_0 are respectively the orthosis inertia, gravity torque, solid friction coefficient, viscous friction coefficient and the stiffness coefficient. We consider the subject in a seated position, and the shank is hanging without any contact with the floor. Flexion and extension movements of the knee joint are studied in this work. The system

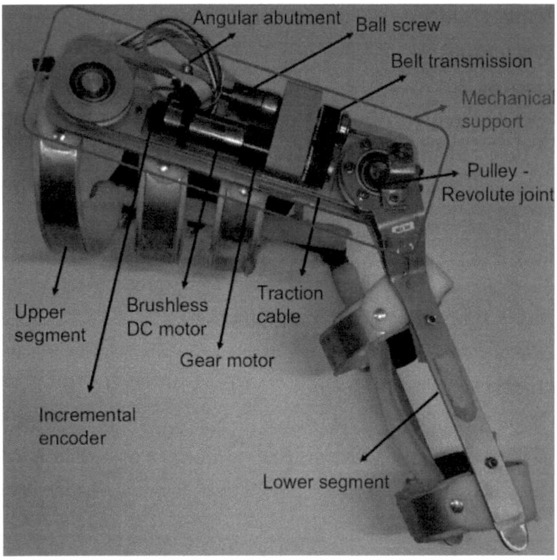

Fig. 1 EICOSI orthosis and its components [12]

tries to track a desired trajectory [13]. By deriving the Lagrangian of the shank-foot-exoskeleton system, the rotational dynamics can be obtained:

$$J\ddot{\theta} = -\tau_g \cos\theta - ASign\dot{\theta} - B\dot{\theta} - K(\theta - \theta_{rest}) + \tau + \tau_h \tag{1}$$

With

$$\begin{aligned} J &= J_s + J_0,\ A = A_s + A_0,\ B = B_s + B_0, \\ \tau_g &= \tau_{gs} + \tau_{g0},\ K = K_s + K_0. \end{aligned} \tag{2}$$

θ is the knee joint angle between the actual position of the shank and the full extension position, $\dot{\theta}$ and $\ddot{\theta}$ are respectively the knee joint angular velocity and acceleration, θ_{rest} is the rest position of the shank. Sign () is a signum function, τ is the control torque and τ_h is the human torque.

Parameters Identification: The identification of the shank-Orthosis parameters was obtained in [13] by separate identification of each component. A nonlinear least square optimization method was used to find the shank and orthosis parameters. θ_{rest} which is the rest position was obtained by means of the incremental encoder then used in the identification process. The Muscular activity of the muscles was monitored through the EMG electrodes. Theses signals are supposed to be null during the rest position. To determine the shank mass and its gravity center, the regression equations of Winter were used considering the subject's weight and height. A passive pendulum test was performed to identify the other parameters [13]. The identified parameters are shown in Tables 1 and 2.

Table 1 Orthosis parameters identification [13]

Parameters	Value ± s.d
Inertia (J_0)	0.0117 ± 3.5238 kg m^2
Solid friction coefficient (A_0)	0.3525 ± 0.2491 N m
Viscous friction coefficient (B_0)	0.6928 ± 0.3811 N m s rad^{-1}
Gravity torque (τ_{g0})	0.2424 ± 0.5518 N m
Stiffness(K_0)	0 N m rad^{-1}

Table 2 Shank parameters identification [13]

Parameters	Value ± s.d
Inertia (J_s)	1.5383 ± 0.7052 kg m^2
Solid friction coefficient (A_s)	0.3975 ± 1.3265 N m
Viscous friction coefficient (B_s)	4.7528 ± 2.4557 N m s rad^{-1}
Gravity torque (τ_{gs})	17.4576 N m
Stiffness(K_s)	0.45 ± 0.5476 N m rad^{-1}

The above parameters were verified using a cross-validation process. A trajectory was considered and the inverse dynamics of the orthosis were computed then the resulting torque was compared to the measured one when the given trajectory was applied.

3 Exoskeleton Control

The following control laws are tested in order to compare different strategies for the orthosis control:

3.1 PID Control

A standard PID controller , also called the three-term controller, has a transfer function that is generally written in the form given bellow:

$$G(s) = K_p + K_i \frac{1}{s} + K_d s \tag{3}$$

where K_p is the proportional gain, K_i is the integral gain and K_d is the derivative gain. A proportional term for an overall control action proportional to the error signal; an integral term to reduce the steady-state errors, and a derivative term to improve the transient response through high-frequency compensation by a differentiator. For optimum performance K_p,K_i and K_d are mutually dependent in tuning [7].

Stability analysis: The action torque of a PID controller may be expressed by :

$$\tau = K_p\tilde{\theta} + K_d\dot{\tilde{\theta}} + K_i\zeta \tag{4}$$

where $\tilde{\theta}$ and $\dot{\tilde{\theta}}$ are respectively the errors between current and desired position and velocity, and ζ is an additional state variable

$$\dot{\zeta} = \tilde{\theta} \tag{5}$$

$$\tilde{\theta} = \theta_d - \theta \tag{6}$$

$$\dot{\tilde{\theta}} = \dot{\theta}_d - \dot{\theta} \tag{7}$$

We consider a free moving orthosis with $\tau_h = 0$.

The closed loop equation obtained by substituting τ may be written as following:

$$\begin{aligned} J\ddot{\theta} + \tau_g \cos\theta + ASign\dot{\theta} + B\dot{\theta} + K(\theta - \theta_{rest}) \\ = K_p\tilde{\theta} + K_d\dot{\tilde{\theta}} + K_i\zeta \\ \dot{\zeta} = \tilde{\theta} \end{aligned} \tag{8}$$

The unique equilibrium of the system is a constant vector :

$$\begin{pmatrix} \zeta_{eq} \\ \tilde{\theta}_{eq} \\ \dot{\tilde{\theta}}_{eq} \end{pmatrix} = \begin{pmatrix} \frac{1}{K_i}\tau_g \cos\theta_d \\ 0 \\ 0 \end{pmatrix} \tag{9}$$

$$\frac{d}{dt}\begin{pmatrix} \zeta \\ \tilde{\theta} \\ \dot{\tilde{\theta}} \end{pmatrix} = \begin{pmatrix} \tilde{\theta} \\ \dot{\tilde{\theta}} \\ \ddot{\theta}_d + \frac{1}{J}(K_p\tilde{\theta} + K_d\dot{\tilde{\theta}} + K_i\zeta - \\ \tau_g \cos\theta - ASign\dot{\theta} - B\dot{\theta} - K(\theta - \theta_{rest})) \end{pmatrix} \tag{10}$$

By operating the following change of variables:

$$\omega = \alpha\zeta - \frac{\alpha}{k_i}\tau_g \cos\theta_d + \tilde{\theta} \tag{11}$$

With $\alpha > 0$ We define the following globally positive definite Lyapunov function:

$$V(\tilde{\theta}, \dot{\theta}, \omega) = \frac{1}{2}[\frac{\omega^2}{\alpha} K_i + \alpha K_d \tilde{\theta}^2 + J\alpha\dot{\theta}^2 - 2\dot{\theta}\tilde{\theta}J]$$
$$+\frac{1}{2}\tilde{\theta}^2[K_p - \frac{1}{\alpha}K_i] + U(\theta_d - \theta) - U(\theta_d) + \tilde{\theta}\tau_g \cos\theta_d \tag{12}$$

where U denote the potential energy of the system and

$$\frac{\partial U}{\partial \theta} = \tau_g \cos\theta \tag{13}$$

α is the positive constant used in the variable chagne in (11) The time derivative of this equation will be :

$$\dot{V}(\tilde{\theta}, \dot{\theta}, \omega) = -\dot{\theta}^2(K_d - \alpha J) - \tilde{\theta}^2(\alpha K_p - K_i)$$
$$- \alpha\tilde{\theta}(\tau_g \cos\theta_d - \tau_g \cos\theta) \tag{14}$$

Using the following proprety related to the Gravitational torques of robotic systems: "*For robots, having exclusively revolute joints, there exists a number* $K_g > 0$ *such as* : $K_g \geq \frac{\partial g(q)}{\partial q}$" [14] We assume: $K_g = \tau_g$ which satisfies the above, and α satisfies the following inequalities:

$$\frac{K_d}{J} > \alpha > \frac{K_i}{K_p - K_g} \tag{15}$$

and

$$K_p > K_g = \tau_g \tag{16}$$

Hense :

$$\begin{gathered} -\dot{\theta}^2(K_d - \alpha J) \leq 0 \\ -\tilde{\theta}^2(\alpha K_p - K_i) \leq 0 \\ -\alpha\tilde{\theta}(\tau_g \cos\theta_d - \tau_g \cos\theta) \leq 0 \quad \forall\theta \in \mathcal{D} \\ \mathcal{D} = [-\frac{\pi}{2}, 0] \end{gathered} \tag{17}$$

It is easily established that, when α is picked to satisfy the condition (15), $V(\tilde{\theta}, \dot{\theta}, \omega)$ is globally positive definite, while the derivative of Lyapunov function (14) is negative semidefinite. If we can find conditions on the gains K_p, $K_i and K_d$ of the PID controller such that the origin of the closed-loop (10) is asymptotically stable, at least for sufficiently small values of the initial states $\omega(0)$, $\tilde{\theta}(0)$ and $\dot{\theta}(0)$, and particularly $\tilde{\theta}(t)$; the state tends asymptotically to zero, then we can conclude that at least locally the position control objective is achieved [14].

3.2 Sliding Mode Control

Sliding mode is one of the commonly used methods to control nonlinear systems. Intuitively the approach is based on the remark that it is easier to control a first order system. A time varying surface is defined as following:

$$s = (\frac{d}{dt} + \lambda)\tilde{\theta} \tag{18}$$

With λ a positive constant [15].
Stability Analysis We consider a free moving orthosis with $\tau_h = 0$ Considering the following Lyapunov function:

$$V = \frac{1}{2}s^2 \tag{19}$$

To satisfy the condition $\dot{V} \leq 0$, the following control torque is proposed:

$$\begin{aligned} \tau =& \tau_g \cos\theta + ASign\dot{\theta} + B\dot{\theta} + K(\theta - \theta_{rest}) \\ &+ J[-\lambda\dot{\theta} + \ddot{\theta}_d + \lambda\dot{\theta}_d + K_{slide}sign(s)] \end{aligned} \tag{20}$$

With K_{slide} a positive gain. The time derivative of the Lyapunov function is :

$$\begin{aligned} \dot{V} &= s\dot{s} = s(\ddot{\tilde{\theta}} + \lambda\dot{\tilde{\theta}}) \\ &= s(\ddot{\theta}_d - \tfrac{1}{J}[-\tau_g \cos\theta - ASign\dot{\theta} - B\dot{\theta} - K(\theta - \theta_{rest}) + \\ &\qquad \tau] + \lambda\dot{\tilde{\theta}} \end{aligned} \tag{21}$$

By replacing τ by its expression in 20, we obtain :

$$\dot{V} = -sk_{slide}Sign(s) \tag{22}$$

Since $s.sign(s) \leq 0\ \forall s$, $\dot{V} \leq 0$ the control torque applied ensure the asymptotic stability of the system [16].

3.3 Adaptive Control

Inspired by the sliding mode approach [8, 15], this model reference control law takes advantages in computing continually the system parameters to achieve better tracking performances of the desired trajectory. Assumption: the current and desired states up to the third derivative are bounded. The time varying surface is defined as following : $s = (\dot{\tilde{\theta}} + \lambda\tilde{\theta})$ with λ a positive scalar The dynamocs of the adaptive law are given:

$$\dot{\hat{J}} = -a_1(\ddot{\theta}_d - \lambda\dot{\tilde{\theta}})s \qquad \dot{\hat{A}} = -a_2 sign(\dot{\theta})s$$
$$\dot{\hat{B}} = -a_3\dot{\theta}s \qquad \dot{\hat{\tau}}_g = -a_4 \cos\theta s \tag{23}$$
$$\dot{\hat{K}} = -a_5(\theta - \theta_{rest})s$$

With $a_i, i \in 1, .., 5$ are positive scaling parameters. Where $\hat{J}$, $\hat{A}$, $\hat{B}$ are the estimated system inertia, solid and viscous friction parameters, gravity torque and stiffness. The control torque is given by the following formula

$$\tau = \hat{J}(\ddot{\theta}_d - \lambda\dot{\tilde{\theta}}) + \hat{A}Sign(\dot{\theta}) + \hat{B}\dot{\theta}$$
$$+\hat{\tau}_g \cos\theta + \hat{K}(\theta - \theta_{rest}) - K_{adapt}s \tag{24}$$

With K_{adapt} a positive gain.
Stability Analysis We consider a free moving orthosis $\tau_h = 0$. The stability of the system when implemented this control law was studied in [8]. We define the positive Lyapunov function:

$$V = \frac{1}{2}Js^2 + \frac{1}{2a_1}\tilde{J}^2 + \frac{1}{2a_2}\tilde{A}^2 + \frac{1}{2a_3}\tilde{B}^2$$
$$+ \frac{1}{2a_4}\tilde{\tau_g}^2 + \frac{1}{2a_5}\tilde{K}^2 + K_{adapt}\tilde{\theta}^2 \tag{25}$$

where: $\tilde{J} = J - \hat{J}$, $\tilde{A} = A - \hat{A}$, $\tilde{B} = B - \hat{B}$, $\tilde{\tau_g} = \tau_g - \hat{\tau}_g$, $\tilde{K} = K = \hat{K}$ The derivative of this Lyapunov function is :

$$\dot{V} = sJ\dot{s} + \frac{1}{a_1}\tilde{J}\dot{\tilde{J}} + \frac{1}{a_2}\tilde{A}\dot{\tilde{A}} + \frac{1}{a_3}\tilde{B}\dot{\tilde{B}} + \frac{1}{a_4}\tilde{\tau}_g\dot{\tilde{\tau}}_g$$
$$+\frac{1}{a_5}\tilde{K}\dot{\tilde{K}} + 2K_{adapt}\lambda\tilde{\theta}\dot{\tilde{\theta}} = -k_{adapt}\dot{\tilde{\theta}}^2 - k_{adapt}\lambda^2\tilde{\theta}^2 \leq 0 \tag{26}$$

All the signals and their derivatives are bounded, hence, the second derivative of the Lyapunov function is bounded which implies that V is uniformly continuous. The state $(\tilde{\theta}, \dot{\tilde{\theta}})$ asymptotically converges to zero with a domain of attraction equal to $[-\pi, \pi]$.

4 Simulation Results

The three control laws are simulated using Matlab/Simulink considering the model of orthosis EICOSI of LISSI-Lab. This control laws are applied to the orthosis in order to track a predefined trajectory taking into account an external human torque. Five scenarios are tested in this simulation (Fig. 2).

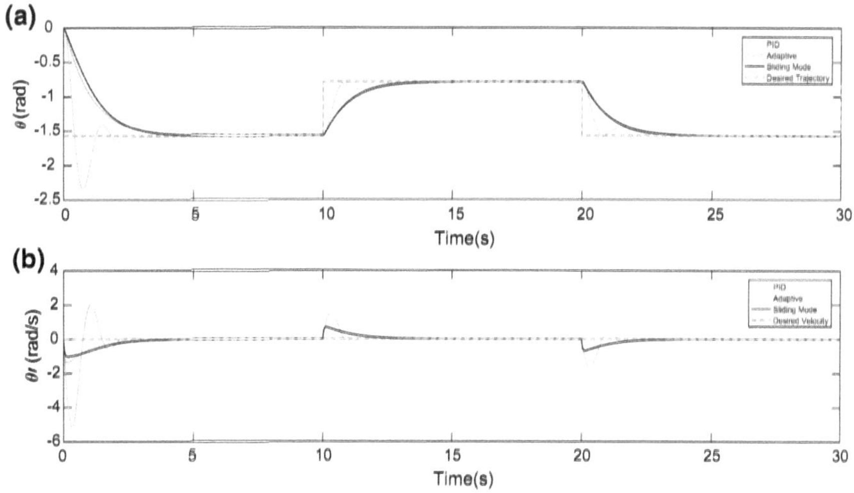

Fig. 2 Flexions and extensions with no human effort: **a** The knee-joint angle. **b** Angular velocity

Scenario 1 The users effort is considered null ($\tau_h = 0$). The control law is tested with a desired trajectory representing flexion and extension movements $[-\frac{\pi}{2}, -\frac{\pi}{4}]$. A white Gaussian noise with null mean and standard deviation of 0.07 N m is added to the torque $\tau_h = 0$ in order to simulate real conditions. For this scenario, a particular attention should be paid to the control torques values (Figs. 3 and 4). For same performances, different torques values are delivered by sliding mode and the adaptive sliding mode controller. These results are a significant indication for the energy

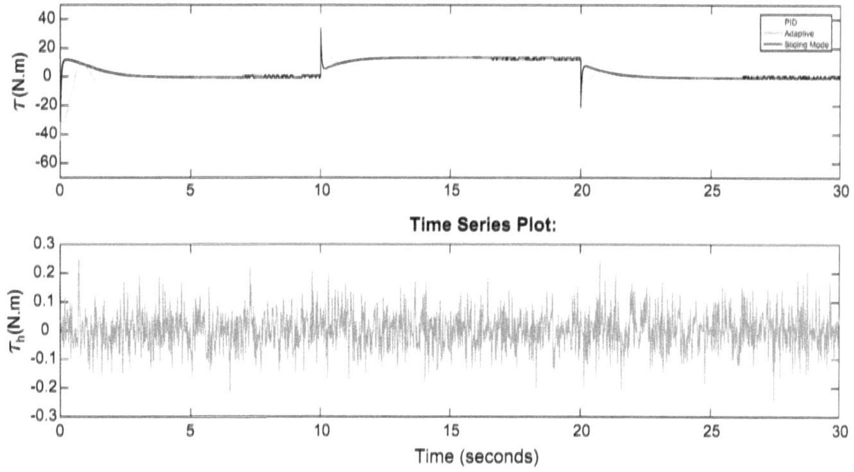

Fig. 3 Flexions and extensions with no human effort: The human and control torques

consumption. Without adaptive mechanism, sliding mode controller consumes less energy. This aspect could be another parameter to be taken in consideration while designing controllers for such nonlinear systems.

Scenario 2 The user's effort is considered null ($\tau_h = 0$). The control law is tested with a sinusoidal desired trajectory $\theta_d = -\frac{1}{4}\sin(2\pi \times 0.2 \times t - \frac{1}{4})$. This simulation put in evidence the inaccuracy of a PID controller. Adding to this, it was proven

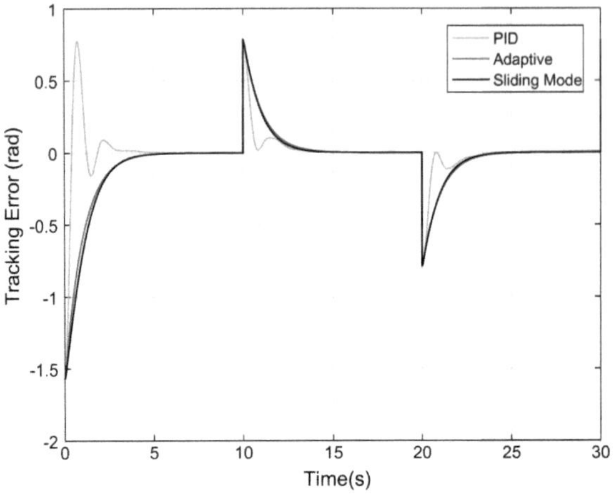

Fig. 4 Flexions and extensions with no human effort: Tracking Error

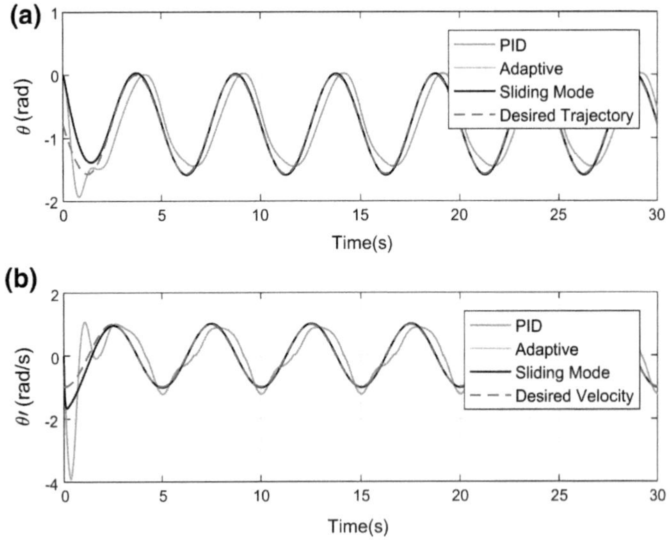

Fig. 5 Sinusoidal trajectory with no human effort: **a** The knee-joint angle. **b** The knee-joint angular velocity

that the human limb movements yield to variations in the model parameters, that a classical PID doesnt compensate. In [17], the authors have modelized the muscular contraction in a viscous environment by a linear damper element that can be described as following:

$$\bar{F}_d^m = \bar{b}_m \bar{v}_m \tag{27}$$

where $\bar{b}_m$ is the normalized damping coefficient and $\bar{v}_m$ is the normalized contraction velocity. In [18], it was put in evidence that there is a correlation between this velocity and the muscle length during muscular contraction. Several works [1, 19] have studied the biomechanical properties related to human movement. Because of these properties, more robust control approaches should be developed to deal with uncertainty and disturbances specific to systems evolving human movement (Figs. 5, 6 and 7).

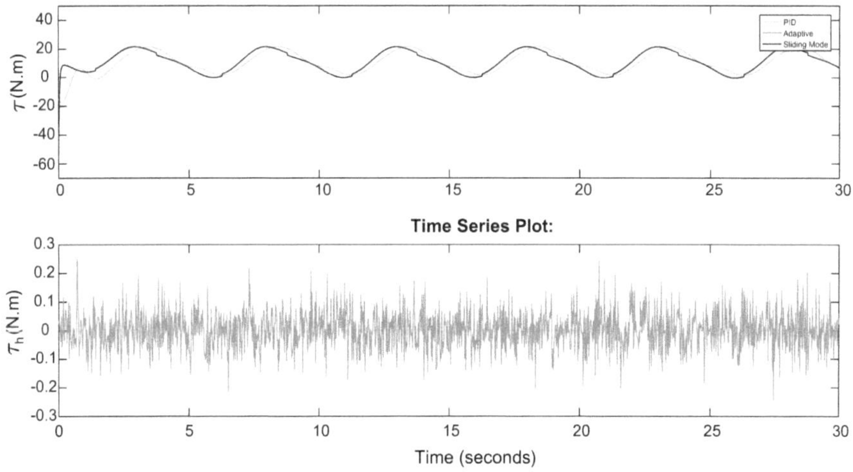

Fig. 6 Sinusoidal trajectory with no human effort: The human and control torques

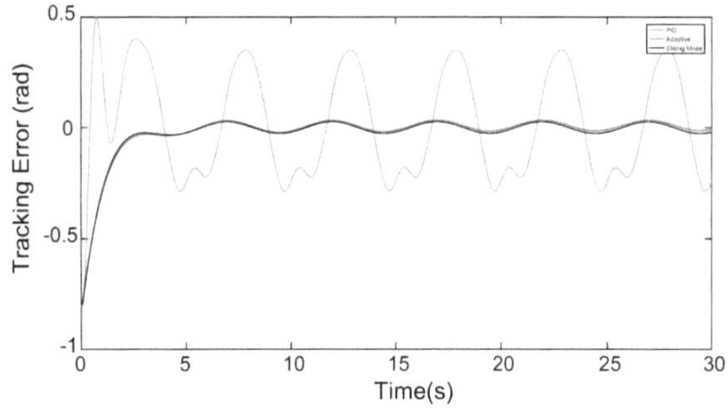

Fig. 7 Sinusoidal trajectory with no human effort: Tracking Error

Scenario 3 The wearer develops an effort to ensure the desired movement ($\tau_h \neq 0$). A sinusoidal signal $\tau_h = 4 \times \sin(2\pi \times 0.2 \times t)$ is added to represent the assisting human effort. The user applies at t = 10 s an assisting torque for 20 s. At t = 30 s, the user stops his effort and the orthosis is driven only by the actuators torque. The control law is tested with a sinusoidal desired trajectory $\theta_d = -\frac{1}{4}\sin(2\pi \times 0.2 \times t - \frac{1}{4})$. The main motivation of this scenario is to observe the control values and the trajectory tracking performances when an assistive human effort is added in an Assistance As Needed context. The control laws are compared in terms of trajectory tracking accuracy and the additional control torque delivered. Figure 8 shows the added human

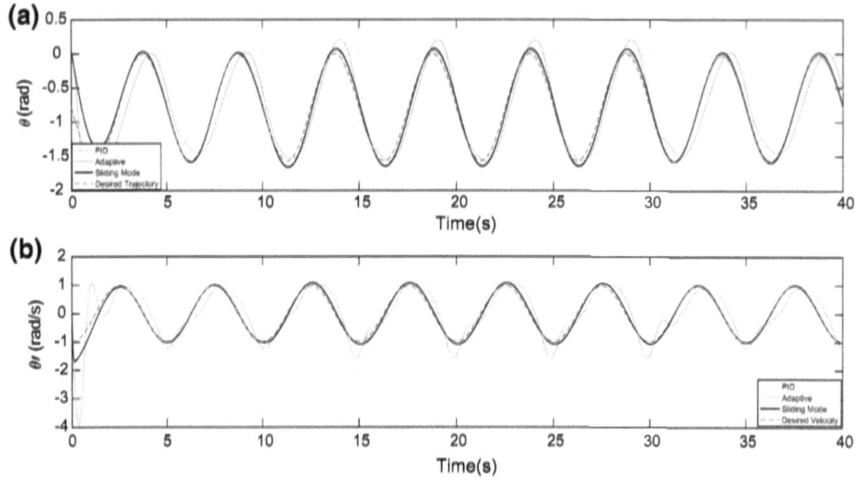

Fig. 8 Sinusoidal trajectory with assisting human effort: **a** The knee-joint angle. **b** The knee-joint angular velocity

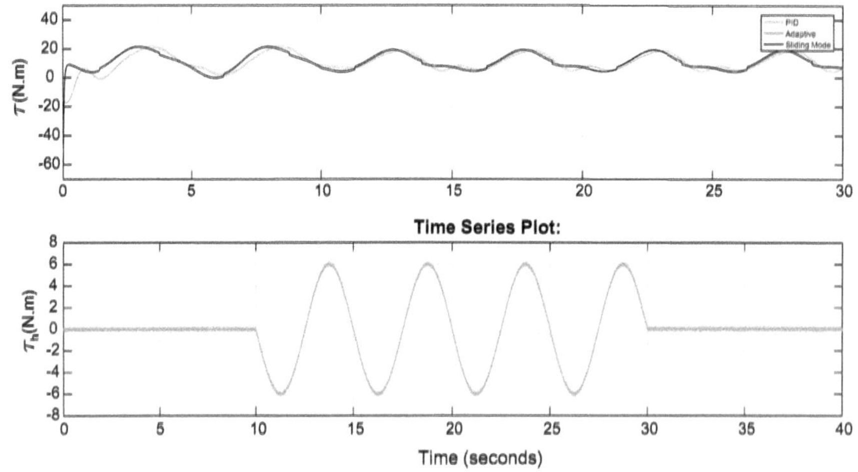

Fig. 9 Sinusoidal trajectory with assisting human effort: The human and control torques

torque results in a small overshoot of the knee-joint angle. For the adaptive controller, the tracking errors were decreasing during the period when the human torque was applied. This is due to the parameters adaptation. A better regulation can be obtained by increasing the parameters convergence gain a_i (Figs. 9 and 10).

Scenario 4 In this scenario, we consider a resistive subject who develops an effort from t = 10 s till t = 30 s, that is opposing to the orthosis movement. A sinusoidal signal $\tau_h = 4 \times \sin(2\pi \times 0.2 \times t)$ is added to represent the resistive human effort. The user applies it at t = 10 s for 20 s. At t = 30 s, the user stops his resistance and the orthosis is driven only by the actuators torque. The control law is tested with a sinusoidal desired trajectory:$\theta_d = -\frac{1}{4}\sin(2\pi \times 0.2 \times t - \frac{1}{4})$. This scenario considers the resistive rehabilitation case. The results show that the Sliding mode controller has better tracking performances in terms of tracking errors. The adaptive

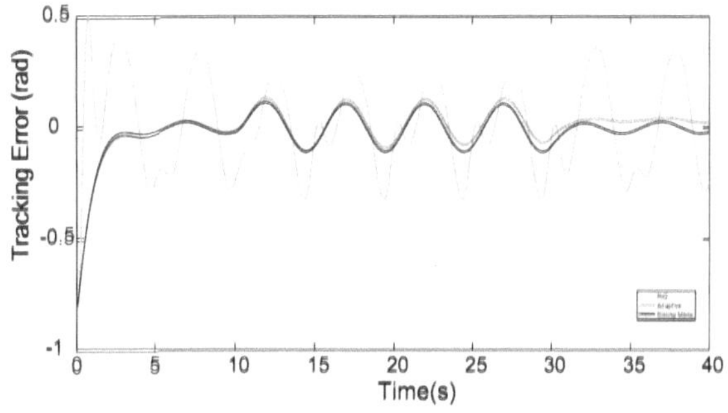

Fig. 10 Sinusoidal trajectory with assisting human effort: Tracking error

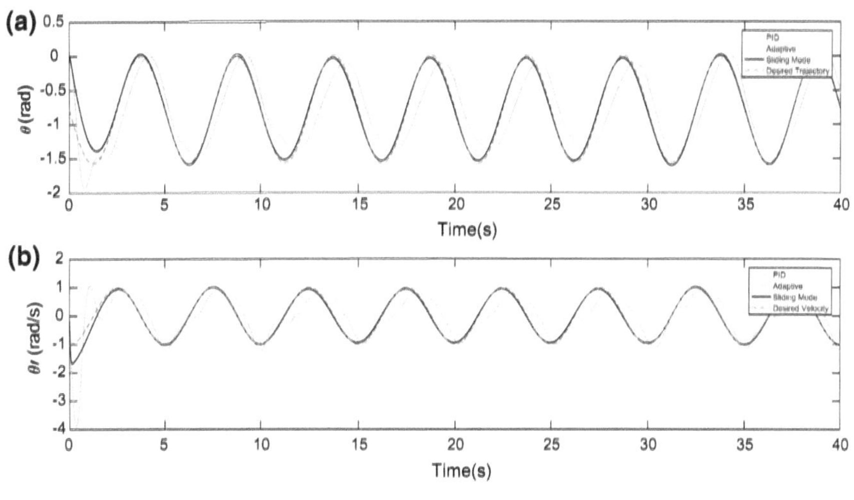

Fig. 11 Sinusoidal trajectory with resisting human effort: **a** The knee-joint angle. **b** The knee-joint angular velocity

controller tends to decrease these errors by adapting the parameters to ensure the better performances (Figs. 11, 12, 13, 14 and 15).

Scenario 5 We consider the second scenario conditions (sinusoidal trajectory). The purpose of this scenario is to evaluate the robustness of the controllers with regard to parameters uncertainties. Bias of 20, 40, 80% of parameters values were introduced to the model parameters. The simulation results show that parameters errors introduce a fixed tracking error. This error is consequently reduced by the adaptation mechanism with the adaptive control Fig. 16. Figure 17 Shows how the error is reduced during the adaptation process. The parameters adapt to ensure a better convergence of the trajectory.

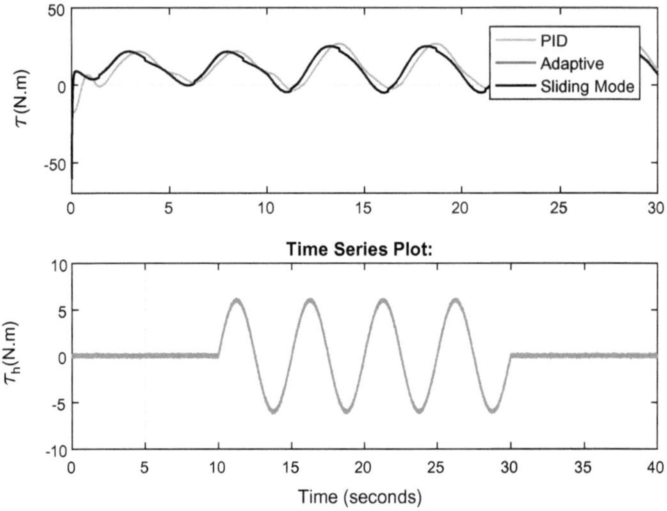

Fig. 12 Sinusoidal trajectory with resisting human effort: The human and control torques

Fig. 13 Sinusoidal trajectory with resisting human effort: Tracking error

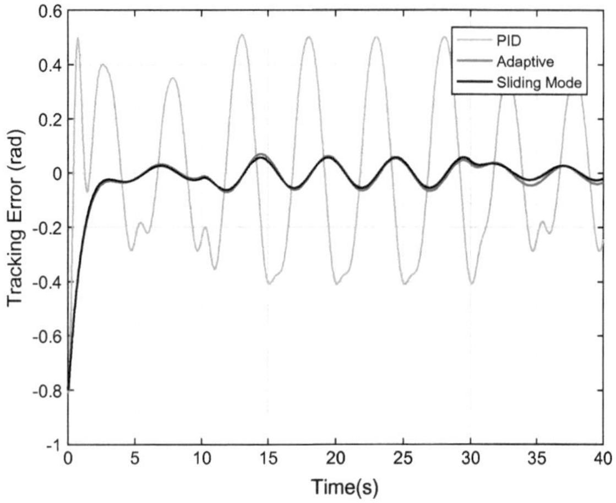

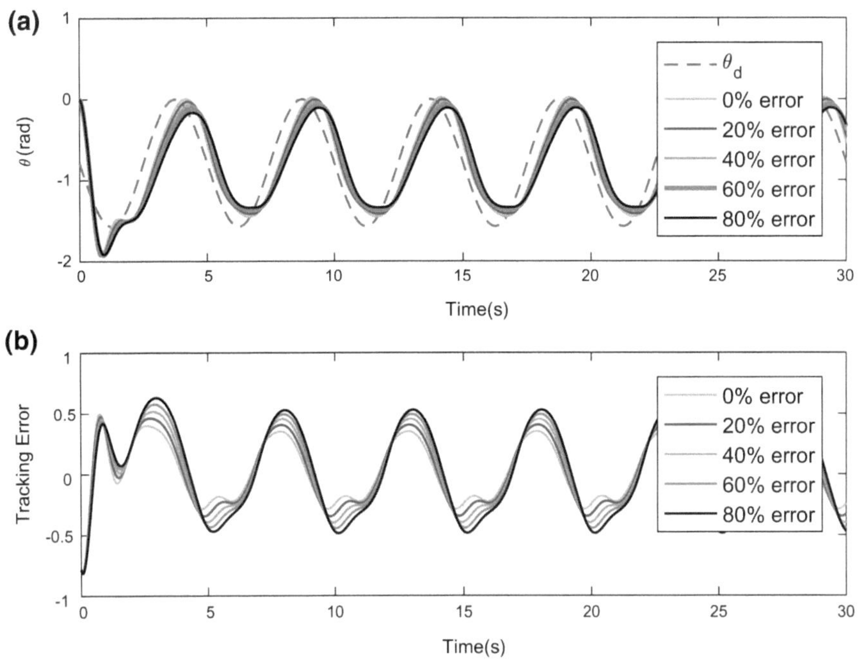

Fig. 14 PID robustness tests: The knee joint angle and tracking error

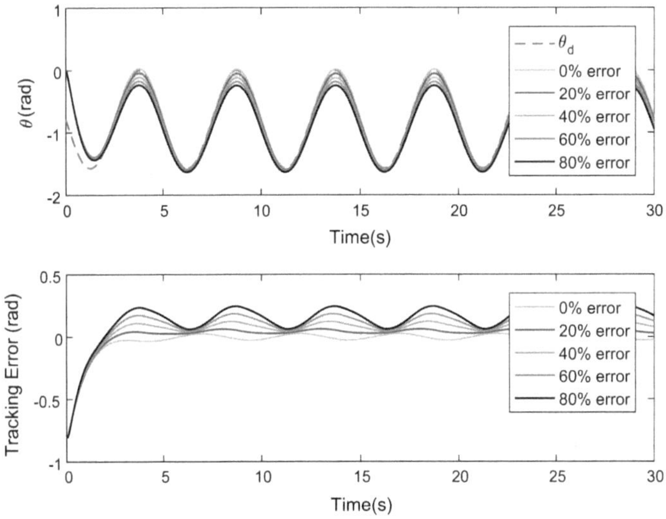

Fig. 15 Sliding mode robustness tests: The knee joint angle and tracking error

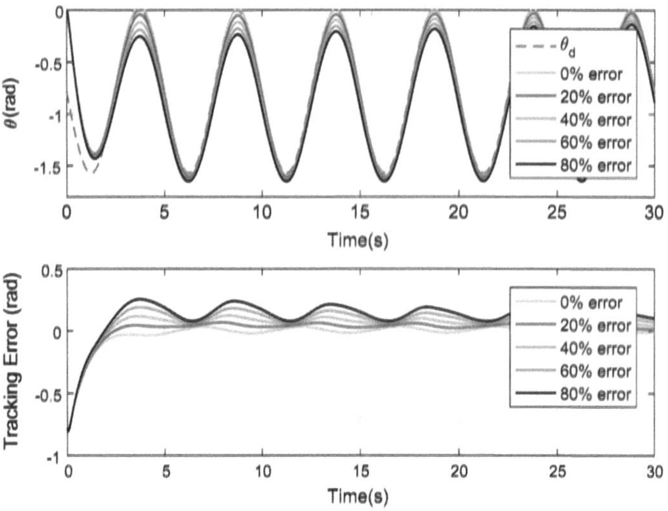

Fig. 16 Adaptive control robustness tests: The knee joint angle and tracking error

Fig. 17 Adaptive control robustness tests: Parameters adaptation

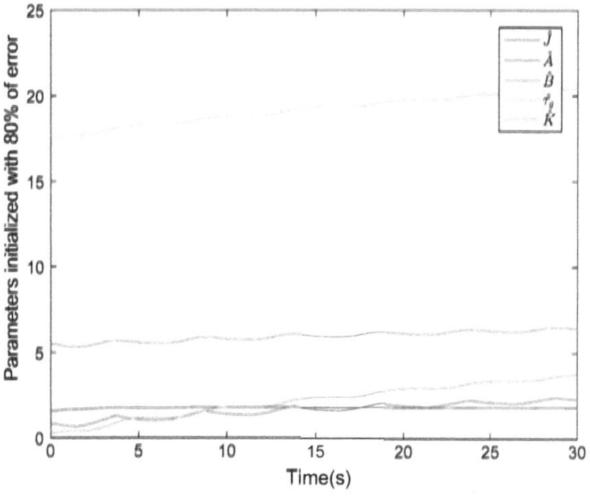

5 Conclusion and Future Work

The main objective of this study is to bring out the benefit of orienting control strategies towards adaptive approach for human assistance and rehabilitation processes to compensate for changes in the human-orthosis dynamics. Three different control laws were tested in different operational scenarios covering three study cases which are: passive rehabilitation, assistance as needed and resistive rehabilitation contexts, a robustness test with regards to parameters uncertainties was performed on the three controllers. The simulation results confirmed that an additional adaptive model based terms including some of the system dynamics can compensate for uncertainties, disturbances and for dynamic changes due to human activity. Future experiments may consolidate those results and demonstrate the relevance of such approaches. This work can be further improved and enriched with real experimentations on human assistance wearable robot.

References

1. Jansen CM, Windau JE, Bonutti PM, Brillhart MV (1996) Treatment of a knee contracture using a knee orthosis incorporating stress-relaxation techniques. Phys Ther 76(2):182–186
2. Rudhe C, Albisser U, Starkey ML, Curt A, Bolliger M (2012) Reliability of movement workspace measurements in a passive arm orthosis used in spinal cord injury rehabilitation. J Neuroeng Rehabil 9(1):37
3. Marchal-Crespo L, Reinkensmeyer DJ (2009) Review of control strategies for robotic movement training after neurologic injury. J Neuroeng Rehabil 6(1):20
4. Buchanan TS, Lloyd DG, Manal K, Besier TF (2004) Neuromusculoskeletal modeling: estimation of muscle forces and joint moments and movements from measurements of neural command. J Appl Biomech 20(4):367–395
5. Fleischer C, Hommel G (2008) A human-exoskeleton interface utilizing electromyography. IEEE Trans Rob 24(4):872–882
6. Ferris DP, Lewis CL (2009) Robotic lower limb exoskeletons using proportional myoelectric control. In: EMBC 2009 annual international conference of the IEEE engineering in medicine and biology society. IEEE, pp 2119–2124
7. Ang KH, Chong G, Li Y (2005) Pid control system analysis, design, and technology. IEEE Trans Control Syst Technol 13(4):559–576
8. Rifai H, Mohammed S, Daachi B, Amirat Y (2012) Adaptive control of a human-driven knee joint orthosis. In: 2012 IEEE International Conference on Robotics and Automation (ICRA). IEEE, pp 2486–2491
9. Blaya JA, Herr H (2004) Adaptive control of a variable-impedance ankle-foot orthosis to assist drop-foot gait. IEEE Trans Neural Syst Rehabil Eng 12(1):24–31
10. Noble JW, Prentice SD (2006) Adaptation to unilateral change in lower limb mechanical properties during human walking. Exp Brain Res 169(4):482–495
11. Kao P-C, Ferris DP (2009) Motor adaptation during dorsiflexion-assisted walking with a powered orthosis. Gait Posture 29(2):230–236
12. Rifaï H, Abdessalem MB, Chemori A, Mohammed S, Amirat Y (2016) Augmented-l adaptive control of an actuated knee joint exoskeleton: from design to real-time experiments. In: 2016 IEEE international conference on robotics and automation (ICRA). IEEE, pp 5708–5714
13. Rifaï H, Mohammed S, Hassani W, Amirat Y (2013) Nested saturation based control of an actuated knee joint orthosis. Mechatronics 23(8):1141–1149
14. Kelly R, Davila VS, Perez JAL (2006) Control of robot manipulators in joint space. Springer Science & Business Media
15. Slotine J-JE, Li W (1987) On the adaptive control of robot manipulators. Int J Robot Res 6(3):49–59
16. Boualbani M, Bousri A (2016) Commande avancée d'un exosquelette. Master's thesis, Université de Montpellier 2, France, Montpellier
17. Thelen DG et al (2003) Adjustment of muscle mechanics model parameters to simulate dynamic contractions in older adults. Trans Am Soc Mech Eng J Biomech Eng 125(1):70–77
18. Hill A (1938) The heat of shortening and the dynamic constants of muscle. Proc R Soc Lond B Biol Sci 126(843):136–195
19. Zajac FE (1989) Muscle and tendon properties models scaling and application to biomechanics and motor. Crit Rev Biomed Eng 17(4):359–411

Robustness Analysis of an Upper Limb Exoskeleton Controlled by Sliding Mode Algorithm

Sana Bembli, Nahla Khraief Haddad and Safya Belghith

Abstract This paper presents a robust sliding mode algorithm developed to control an exoskeleton used for rehabilitation of the upper limb. The considered system is a robot with three degrees of freedom controlling the flexion/ extension movement of the shoulder, the elbow and the wrist. A Monte Carlo simulation is done to analyze the robustness of the controller against matched and unmatched disturbances. Simulation results are provided to prove the performances and the effectiveness of the sliding mode algorithm face to disturbances.

Keywords Upper limb exoskeleton · Matched and unmatched disturbances
Sliding mode · Monte carlo

1 Introduction

Exoskeleton is a mechatronic system placed on the user's body and acts as amplifiers that augment, reinforce or restore human performance. Robotics have naturally emerged in the field of upper/lower limb rehabilitation in the 1960s as an evolution of existing mechanical devices and in response to a need for improved quality of therapies.

The exoskeleton is an articulated mechanical structure made up of the various components such as sensors, actuators and control unit which performs the acquisition and processing of the information delivered by the sensors and of controlling the actuators according to Control laws guaranteeing the effectiveness of the assistance movement and the stability of the system.

Exoskeleton robots are used in different fields of applications. In the context of military applications, these robots are used in order to increase the physical endurance of soldiers and help them lift heavy loads. We find the development of Harvard's

S. Bembli (✉) · N. K. Haddad · S. Belghith
RISC Laboratory, National Engineering School of Tunis, University of Tunis El Manar, Tunis, Tunisia
e-mail: Bembli.sana@hotmail.fr

R. Rizk and M. Awad (eds.), *Mechanism, Machine, Robotics and Mechatronics Sciences*, Mechanisms and Machine Science 58,
https://doi.org/10.1007/978-3-319-89911-4_8

exoskeleton by researchers from Harvard University's Wyss Institute for Biologically Inspired Engineering which allows soldiers to walk longer distances carrying heavy loads with less effort, while also minimizing risk of injury [3]. Around 2010, two major exoskeleton projects for the military were brought to the public's attention: the HULC (Human Universal Load Carrier) developed by Ekso Bionics and Lockheed Martin, and the XOS as well as XOS2 developed by Sarcos and Raytheon. Both projects were full body suits for soldiers' mobility augmentation. We find also the assistance field where exoskeletons are used by dependent persons in their daily physical activities.

In the medical context, while ability to move is very necessary to ensure basic activities of daily living and the number of hemiplegic people is constantly increasing (Referring to the World Health Organization Mackay and Mensah in [1]), the development of robotic exoskeletons, which are systems in physical interaction with the human being used in order to help the patient to realize his movement and to improve more comfort, becomes a powerful solution. In such context, we find the development of "SUEFUL-7" exoskeleton by Saga University used to control all the axes of the upper limb [4]. The Space Systems Laboratory in collaboration with the Georgetown University Imaging Science and Informations Systems have designed an exoskeleton which allows an adaptation to the complex of the shoulder in order to propose a larger workspace [2].

The interaction of the exoskeleton with a human subject must be sufficiently neat to respect the requirements of the human motor control and allow the setting up of controllers dedicated to rehabilitation. In order to follow the movements of a healthy user and augmenting his/her physical capabilities for specific tasks in a relatively safe and transparent way, many control algorithms were developed. In literature, many techniques have been proposed to control an exoskeleton-upper limb system which is characterized by a complexity of the dynamic. Referring to [6], we find the development of several control modes, such as universal approximations of fuzzy logic or neural networks approaches which require offline learning to avoid undesirable behavior of the robot. According to Ref. [5], the authors use sliding mode to control the upper limb exoskeleton. This controller has largely proved its effectiveness through several theoretical studies. There is the development of a mixed force and position controller used by the author in [7].

These controllers are obtained at the cost of certain disadvantages like the performance when tracking desired trajectories and robustness in the presence of disturbances.

Like any robotic system, exoskeletons suffer from different components of uncertainties. We find the most important source of uncertainty is the external interaction forces on the suspended body, which are generally unknown. So, it is necessary to study the stability and the robustness of the considered system face to these uncertainties.

In this context, the contribution of this paper is to study the robustness of the upper limb exoskeleton controlled by the sliding mode law against matched and unmatched disturbances. To do this, a Monte Carlo simulation, which is a probabilistic technique based on the use of a large number of random disturbances, is realized.

The paper is organized as follows: Sect. 2 deals with the problem formulation and the sliding mode controller used. Section 3 describes the robustness analysis and the stability studies of the system affected by matched and unmatched disturbances. In Sect. 4, simulation results and discussions are given. Finally, Sect. 5 is reserved for the conclusion and future work.

2 Modeling and Control of an Upper Limb Exoskeleton

2.1 Problem Formulation

The objective of the proposed sliding mode law is to operate the upper limb exoskeleton in order to help the patient to make desired movements. The proposed system treats the actuated joints shoulder, elbow and wrist.

To test the control law applied to our system, sinusoidal form trajectories were chosen. Generally, these trajectories are defined by a doctor.

2.2 Dynamic Model

The Lagrange method is often used to obtain the differential equations of the manipulator motion. Like any method, it rests on the laws of mechanics and energy. The dynamic equation can usually be written as follows

$$\Gamma_i = \frac{d}{dt}\left(\frac{\partial L}{\partial \dot{q}i}\right) - \frac{\partial L}{\partial qi} \tag{1}$$

Where Γ_i is the generalized force associated with the articulation variables q_i and L is the Lagrangian of the mechanical system.

Referring to Lagrange equation Eq. 1, the dynamic model of an exoskeleton having three degrees of freedom (DoF), presented by Fig. 1. is given by the following equation

$$M(q)\ddot{q} + C(q, \dot{q})\dot{q} + G(q) + \underbrace{f_v \dot{q} + k_i \operatorname{sign}(\dot{q}_i)}_{F(q,\dot{q})} = \tau \tag{2}$$

$$M(q)\ddot{q} + C(q, \dot{q})\dot{q} + G(q) + F(q, \dot{q}) = \tau \tag{3}$$

Where $q \in \mathbb{R}^n$ is the joint positions vector; $\dot{q} \in \mathbb{R}^n$ is the joint velocities vector; $\ddot{q} \in \mathbb{R}^n$ is the joint accelerations vector; $M(q) \in \mathbb{R}^{nxn}$ is the inertia matrix; $C(q, \dot{q}) \in \mathbb{R}^{nxn}$ is the Coriolis matrix, $G(q) \in \mathbb{R}^n$ is the gravitational vector; $F(q, \dot{q}) \in \mathbb{R}^n$ is the friction force and $\tau \in \mathbb{R}^n$ is the control vector.

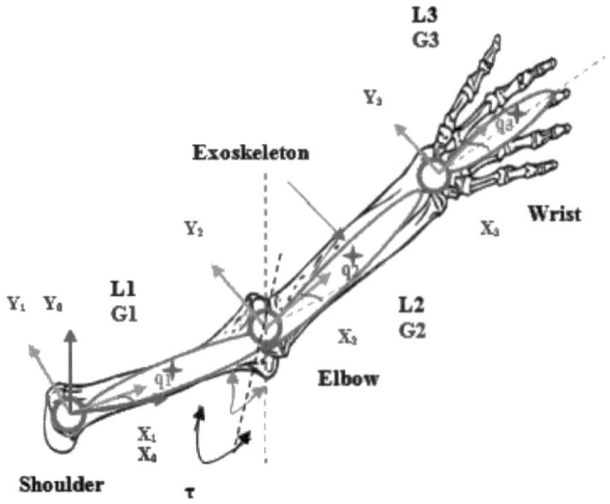

Fig. 1 General configuration of a 3 DoF exoskeleton

We synthesize then the sliding mode algorithm used to control the exoskeleton in order to follow the desired trajectories.

2.3 *Sliding Mode Approximation*

The main objective of this algorithm is to develop a control law U (t) which ensures the achievement and the upkeep of the sliding mode S = 0.

To ensure the convergence of the variable to its desired value, we define the sliding surface by

$$S = \lambda e + \dot{e} \tag{4}$$

Where $e = q_d - q$ presents the tracking error and λ is the vector of parameters for setting the discontinuous control.

The system checked the tracking error when the sliding surface S = 0 is reached. This error is represented by the following equation

$$\lambda e + \dot{e} = 0 \tag{5}$$

We can rewrite the dynamic model of the system presented by Eq. 3 in the following form

$$\ddot{q} = M^{-1}(q)[\tau - C(q, \dot{q})\dot{q} - G(q) - F(q, \dot{q})] \tag{6}$$

Remark: The condition S = 0 means that the velocity converges exponentially towards its reference. Calculating the derivative of S with respect to time

$$\dot{S} = \lambda\,\dot{e} + \ddot{e} \tag{7}$$

We define the sliding mode control applied to the robot with the following equation

$$U(t) = U_{eq} + U_n \tag{8}$$

With U_{eq} corresponds to the equivalent command proposed by Filipov and U_n is determined to check the convergence condition.

To calculate U_{eq}, it is necessary that $\dot{S}=0$, which give

$$\lambda\,e + \ddot{e} = \lambda\,\dot{e} + (\ddot{q}_d - \ddot{q}) = 0 \tag{9}$$

$$\ddot{q} = \lambda\,\dot{e} + \ddot{q}_d \tag{10}$$

Replacing $\ddot{q}$ by its expression in Eq. 3, we get

$$U_{eq} = M(\lambda\,\dot{e} + \ddot{q}_d) + G(q) + C\,(q, \dot{q})\,\dot{q} + F\,(q, \dot{q}) \tag{11}$$

The main purpose of this command is to check the attractiveness conditions.

The commutation control U_n is expressed by

$$U_n = -k\,\mathrm{sign}\,(S)$$

Where k is the gain chosen to guarantee stability, speed and to overcome external disturbances that may affect the system and the "–" symbol indicates the point of operation to the surface at any time of operation.

$$U_n = -k\,\mathrm{sign}\,(\lambda\,e + \dot{e}) \tag{12}$$

Referring to Eqs. 11 and 12, the sliding mode command is given by the following equation

$$\begin{aligned} U = {} & M(q)\,(\lambda\,\dot{e} + \ddot{q}_d) + G(q) + C\,(q, \dot{q})\,\dot{q} + F\,(q, \dot{q}) \\ & - (k\,\mathrm{sign}\,(\lambda\,e + \dot{e})) \end{aligned} \tag{13}$$

We pass then to study the robustness of the exoskeleton controlled by sliding mode law face to disturbances.

3 Robustness Analysis

In order to test the performance of the proposed sliding mode algorithm, we used Monte Carlo simulation [8, 9] and we applied different disturbances to our system.

The dynamical equation of the exoskeleton in the presence of uncertainties can be rewritten as follow

$$\ddot{q} = \mathrm{f}(\mathrm{q}, \dot{\mathrm{q}}, t) + \mathrm{g}(\mathrm{q})(\mathrm{U(t)} + \delta_1) + \delta_2 \tag{14}$$

With $U = \tau$, $g(q) = M^{-1}(q)$ is the input function, δ_1 is the matched disturbance, δ_2 is the unmatched disturbance, and $f(q, \dot{q}, t)$ is given by

$$\mathrm{f}(\mathrm{q}, \dot{\mathrm{q}}, \mathrm{t}) = -\mathrm{M}(\mathrm{q})^{-1}[\mathrm{C}(\mathrm{q}, \dot{\mathrm{q}})\dot{\mathrm{q}} + \mathrm{G}(\mathrm{q}) + \mathrm{F}(\mathrm{q}, \dot{\mathrm{q}})] \tag{15}$$

We study then the stability of the system face to these disturbances applied.

3.1 *Stability Vis-a-Vis Matched Disturbances*

We consider the particular case of a non-linear system in the presence of matched disturbances; the dynamic model is given by the following differential equation

$$\ddot{q} = \mathrm{f}(\mathrm{q}, \dot{\mathrm{q}}) + \mathrm{B}(\mathrm{q}, \dot{\mathrm{q}})\left[\mathrm{U} + \delta_1(\mathrm{q})\right] \tag{16}$$

Where f and B are two nonlinear functions and u is the system input. In this case, $\delta_2 = 0$.

To study the stability of the exoskeleton, we chose the following candidate Lyapunov function is

$$\mathrm{V} = \frac{1}{2}\mathrm{S}^2 \tag{17}$$

Where S is the sliding surface.

V is semi-defined and positive: $\forall$ S, $S^2 > 0$. Thus $\forall$ S, $V \geq 0$. We calculate then $\dot{V}$.

$$\dot{V} = \mathrm{S}\dot{\mathrm{S}} = \mathrm{S}(\lambda\,\dot{\mathrm{e}} + \ddot{\mathrm{e}}) \tag{18}$$

The derivative of the sliding surface is given by

$$\begin{aligned} \mathrm{S} &= \lambda\,\dot{\mathrm{e}} + \ddot{\mathrm{e}} = \lambda\,\dot{e} - \ddot{q}_{\mathrm{d}} + \ddot{q} \\ \dot{\mathrm{S}} &= \lambda\,\dot{\mathrm{e}} + \ddot{\mathrm{e}} = \lambda\,\dot{\mathrm{e}} - \ddot{q}_{\mathrm{d}} + \mathrm{f}(\mathrm{q}, \dot{\mathrm{q}}) + \mathrm{B}(\mathrm{q}, \dot{\mathrm{q}})\ [\mathrm{U} + \delta_1(\mathrm{q})] \end{aligned} \tag{19}$$

To satisfy the condition $\dot{V} < 0$, we chose $\dot{S}$ as follows

$$\dot{\mathrm{S}} = -\mathrm{K}\,\mathrm{sign}\,(\mathrm{S})$$

Thus

$$-K\,\mathrm{sign}(S) = \lambda\dot{e} + \ddot{e} = \lambda\dot{e} - \ddot{q}_d + f(q,\dot{q}) + B(q,\dot{q})\,[U + \delta_1(q)] \tag{20}$$

We deduce then the following control law which will allow the state vector to reach the sliding surface and to converge towards the desired trajectory

$$U = B^{-1}[\ddot{q}_d - \lambda\dot{e} - f(q,\dot{q})] - \delta_1(q) - B^{-1}\,\mathrm{sign}\,(S) \tag{21}$$

$$U = U_{eq} + U_n$$

With :

- $U_{eq} = B^{-1}[\ddot{q}_d - \lambda\dot{e} = f(q,\dot{q})] - \delta_1(q)$
- $U_n = -B^{-1}\,K\,\mathrm{sign}\,(S)$

$\dot{V}$ is given by the following expression

$$\dot{V} = \dot{S}\,S = S[\lambda\dot{e} - \ddot{q}_d + f(q,\dot{q})] + B(q,\dot{q})\,(U + \delta_1(q)]$$

$\dot{V}$ is negative if and only if

$$\lambda\dot{e} - \ddot{q}_d + f(q,\dot{q}) + B(q,\dot{q})[U + \delta_1(q)] \quad \begin{cases} < 0 \text{ for } S > 0 \\ = 0 \text{ for } S = 0 \\ > 0 \text{ for } S < 0 \end{cases}$$

The first derivative of v can be rewritten as

$$\begin{aligned} \dot{V} &= S[\lambda\dot{e} - \ddot{q}_d + f(q,\dot{q})] + B(q,\dot{q})\,\delta_1(q)] + S\,B\,(q,\dot{q}) \\ &\quad (B^{-1}[\ddot{q}_d - \lambda\dot{e} - f(q,\dot{q})] - \delta_1(q) - B^{-1}\,K\,\mathrm{sign}\,(S)) \\ \dot{V} &= -S\,K\ \mathrm{sign}\,(S) < 0 \end{aligned} \tag{22}$$

As the term *-KS sign (S)* <0 because $K \geq 0$ and since the function *sign* is constant by pieces so *S sign (S)* $= +1 > 0$, $\forall S$. Then $\dot{V}$ is semi negative.

As $V \geq 0$ and $\dot{V} \leq 0$, the system is then asymptotically stable.

3.2 Stability Vis-a-Vis Unmatched Disturbances

By applying unmatched uncertainties to the exoskeleton, the dynamic equation will be rewritten as follow

$$\ddot{q} = f(q,\dot{q},t) + B(q,\dot{q})U(t) + \delta_2 \tag{23}$$

Where f and B are two nonlinear functions and u is the system input. In this case, $\delta_1 = 0$.

To study the stability of the system in this case, we chose the same Lyapunov function used in the case of matched disturbances presented by Eq. 17.

Since $S^2 > 0$, $\forall S$ so V is semi-define and positive. Demonstrate then that $\dot{V} < 0$.

Referring to Eq. 23, the derivative of S can be expressed by

$$\dot{S} = \lambda \dot{e} + \ddot{e} = \lambda \dot{e} - \ddot{q}_d + f(q, \dot{q}) + B(q, \dot{q}) U + \delta_2(q) \tag{24}$$

Thus, we can write

$$-K sign(S) = \lambda \dot{e} + \ddot{e} = \lambda \dot{e} - \ddot{q}_d + f(q, \dot{q}) + B(q, \dot{q}) U + \delta_2(q) \tag{25}$$

By isolating u, we deduce the following control law

$$U = B^{-1}[\ddot{q}_d - \lambda \dot{e} - f(q, \dot{q}) - \delta_2(q)] - B^{-1} K \, sign\,(S) \tag{26}$$

$$U = U_{eq} + U_n$$

With:

- $U_{eq} = B^{-1}[\ddot{q}_d - \lambda \dot{e} - f(q, \dot{q}) - \delta_2(q)]$
- $U_n = -B^{-1} K \, sign\,(S)$

$\dot{V}$ can be rewritten as

$$\dot{V} = \dot{S}\, S = S[\lambda \dot{e} - \ddot{q}_d + f(q, \dot{q}) + B(q, \dot{q})\, U + \delta_2(q)]$$

$\dot{V}$ is negative if and only if

$$\lambda \dot{e} - \ddot{q}_d + f(q, \dot{q})] + B(q, \dot{q})\, U + \delta_2(q) \quad \begin{cases} < 0 \text{ for } S > 0 \\ = 0 \text{ for } S = 0 \\ > 0 \text{ for } S < 0 \end{cases}$$

We can write $\dot{V}$ in the following form

$$\begin{aligned} \dot{V} &= S[\lambda \dot{e} - \ddot{q}_d + f(q, \dot{q})] + \delta_2(q)] + S\, B\, (q, \dot{q}) \\ &\quad (B^{-1}[\ddot{q}_d - \lambda \dot{e} - f(q, \dot{q})] - \delta_2(q)] - B^{-1} K \, sign\,(S)) \\ \dot{V} &= -S\, K \, sign\,(S) < 0 \end{aligned} \tag{27}$$

$K \geq 0$ and since the function *sign* is constant by pieces so *S sign (S)* = +1 > 0, $\forall S$. Thus, the term *-KS sign (S)* < 0 which implies $\dot{V}$ is semi negative.

As $V \geq 0$ and $\dot{V} \leq 0$, the system is then asymptotically stable.

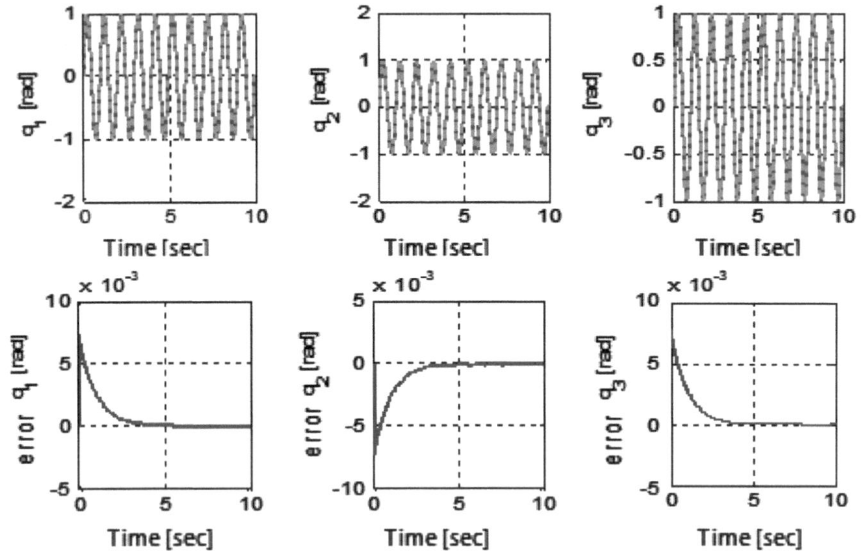

Fig. 2 Simulation results of the positions and errors tracking trajectories of the joints q_1, q_2 and q_3 using sliding mode controller in the nominal case

Table 1 Simulation parameters

Joint	1	2	3
Masses (Kg)	2.75	2.5	1.25
Lengths (m)	0.6	0.55	0.34
Solid friction coefficient (N.m)	0.042	0.03	0.019

4 Simulations and Results

To study the efficiency of the sliding mode controller against disturbances, some tests are developed using Monte Carlo simulation. To conduct to this method, we start by identifying the type of the uncertainties distributions applied to the input system. In our case, we used uniform random distribution applied in the interval [0.01; 1.49] at t=0.2 s to test the robustness of our system face to matched and unmatched disturbances.

Figures 2 and 3 present respectively the position and velocity tracking of desired trajectories and the errors in the nominal case while Figs. 4 and 5 describe the response of the system when applying disturbances.

From the results of these figures, it is interesting to conclude that we get a good tracking of desired trajectories in position and velocity, when controlling with the sliding mode law, despite the existence of disturbances (Table 1).

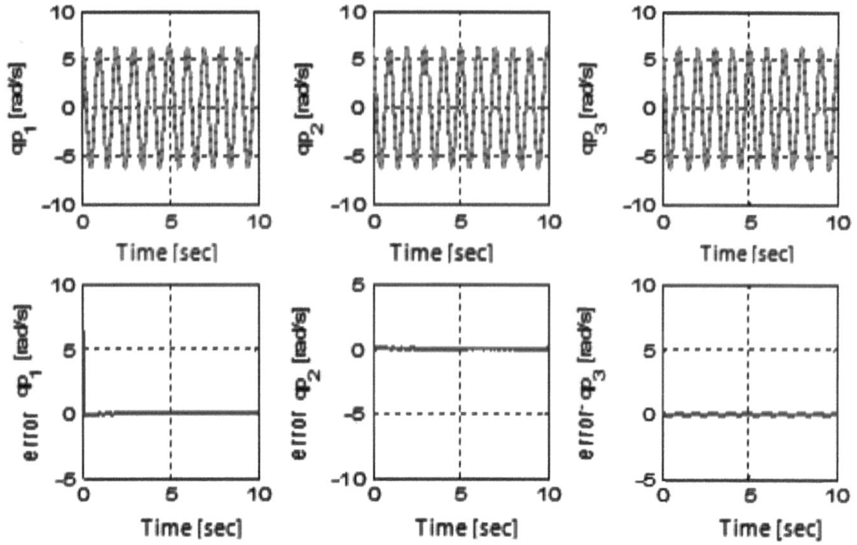

Fig. 3 Simulation results of the velocities and errors tracking trajectories of the joints q_1, q_2 and q_3 using sliding mode controller in the nominal case

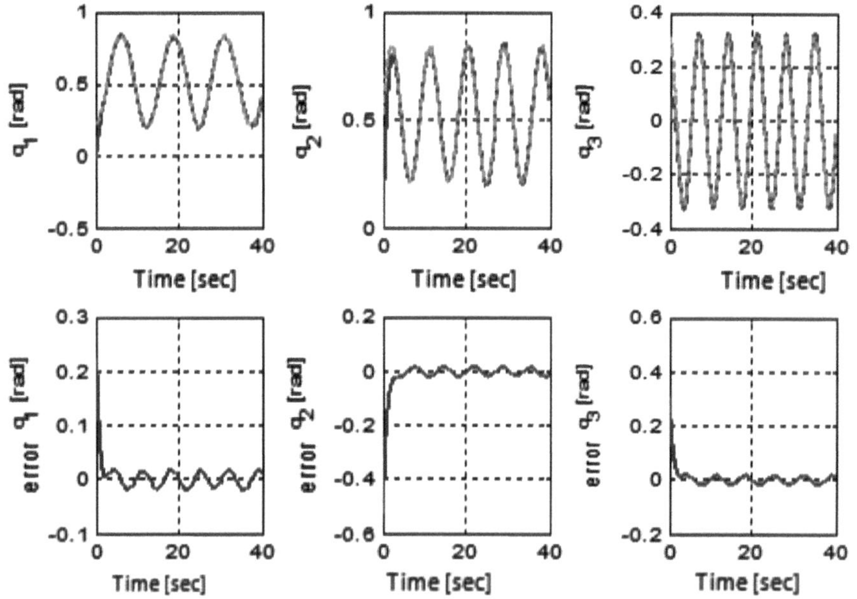

Fig. 4 Simulation results of the positions and errors tracking trajectories of the joints q_1, q_2 and q_3 using sliding mode controller when applying disturbances

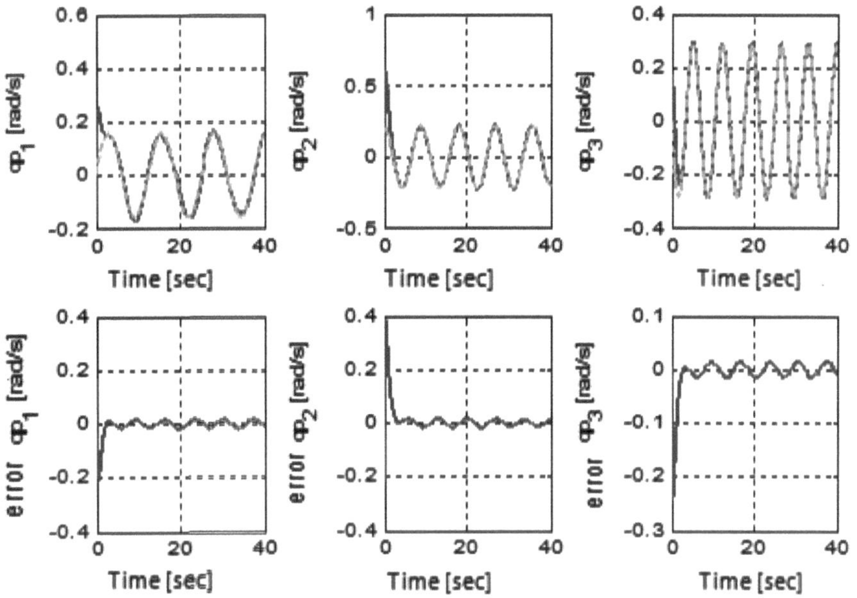

Fig. 5 Simulation results of the velocities and errors tracking trajectories of the joints q_1, q_2 and q_3 using sliding mode controller when applying disturbances

Fig. 6 Calculation of the RMS of the three joints q_1, q_2 and q_3 when tracking the desired trajectories in positions

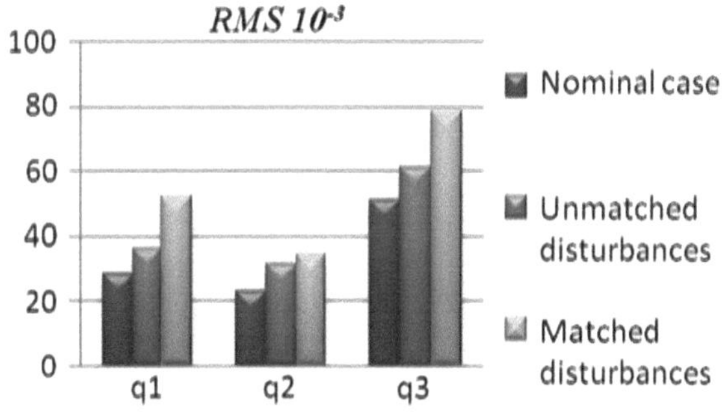

The desired trajectories are given by $q_1 = \sin(2*pi*t)$, $q_2 = \sin(2*pi*t)$ and $q_3 = \sin(2*pi*t)$. The initial conditions of the real trajectories are $q(0) = [0; pi/4; pi/4]^T$ and $\dot{q}(0) = [0; 0; 0]^T$.

To perform the tests, the nominal case of the sliding mode controller was compared to the cases of applying matched and unmatched disturbances. Table 2 and Figs. 6 and 7 present some statistics of the tracking recorded errors. We calculated the Root-Mean-Square (RMS), the mean (Mean) and the standard deviation (Std).

The RMS is given by the following expression

$$X_{RMS} = \sqrt{\frac{1}{N}\sum_{n=1}^{N} |Xn|^2} \quad (28)$$

Table 2 Summary of the results of the Monte Carlo simulation. RMS, mean error and standard deviation calculation for each articulation q_1, q_2 and q_3 using the sliding mode controller in case of tracking the desired trajectories in positions and velocities in the nominal case and when applying matched and unmatched disturbances

		Nominal case			Matched disturbances			Unmatched disturbances		
		q_1	q_2	q_3	q_1	q_2	q_3	q_1	q_2	q_3
Position error	Mean 10^{-3}	5.3	0.54	1.2	38	19.1	34	6.2	0.75	2.4
	RMS 10^{-3}	28	22.7	51	52	34	78	36	31	61
	Std 10^{-3}	28.1	22.7	50	42	46	67	37	32	58
Velocity error	Mean 10^{-3}	2.01	1.8	0.73	36	25	21	2.9	2.01	0.58
	RMS 10^{-3}	31	30	40	31	32.6	40.5	31.9	30	40
	Std 10^{-3}	31.4	30.1	40.02	31	30	40.4	31.3	30.1	40.05

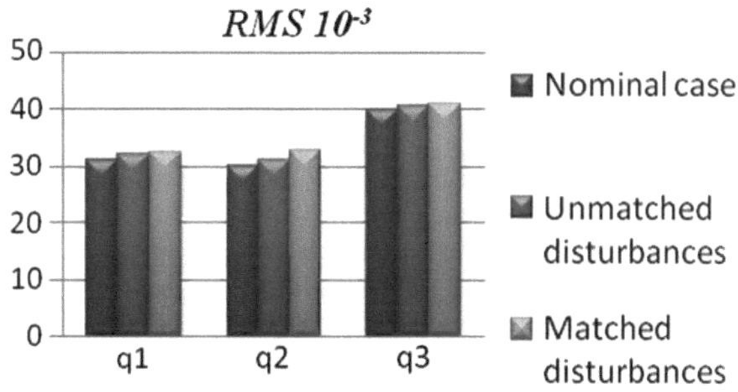

Fig. 7 Calculation of the RMS of the three joints q_1, q_2 and q_3 when tracking the desired trajectories in velocities

The Std is expressed by

$$\sigma_x = \sqrt{E\left[x - E[x])^2\right]} = \sqrt{E\left[x^2\right] - E[x]^2} \tag{29}$$

And the sample mean is defined as

$$\bar{\theta} = \frac{1}{m}\sum\nolimits_{i=1}^{m} \theta_i \tag{30}$$

In the course of our work, we choose the RMS as a criterion of performance and robustness against matched and unmatched disturbances to prove the efficiency of the sliding mode algorithm.

Referring to the results of Figs. 6 and 7 which present the calculated RMS, we can clearly note the performances when tracking the desired trajectories and the robustness against matched and unmatched disturbances of the sliding mode controller.

5 Conclusion

This paper deals with the development of a sliding mode algorithm used to control a three degree of freedom upper limb exoskeleton. A dynamical model of the robot and a stability proof were developed. A Monte Carlo simulation was realized to analyze the robustness of the controller against disturbances. Referring to the simulation results, an analysis of the response of the system affected by matched and unmatched disturbances was done in order to prove significantly the robustness and the effectiveness of the controller. As a future work, a real exoskeleton-upper limb interaction will be studied. A development and a control of this system will be realized by applying trajectories taken from healthy arms.

References

1. Kittel-Ouimet T (2012) Commande d'un bras exosquelette robotique à sept degrés de liberté, memory presented at the school of superior technology Montréal, pp 1–5
2. Naidu D, Stopforth R, Bright G, Davrajh S (2012) A portable passive physiotherapeutic exoskeleton. Int J Advan Robotic Syst
3. Atherton KD (2015) DARPA tests exoskeletons on soldiers. 22 Sept 2015
4. Gopura RARC, Student Member, IEEE, Kiguchi K, Member, IEEE, Li Y (2009) SUEFUL-7: a 7dof upper-limb exoskeleton robot with muscle-model-oriented emg-based control. In: The 2009 IEEE/RSJ international conference on intelligent robots and systems
5. Bouteraa Y, Abdallah IB (2017) Estimated model-based sliding mode controller for an active exoskeleton robot, chapter 10. In: Applications of sliding mode control, decision and control, pp 175–189
6. Su H, Li Z, Li G, Yang C (2013) EMG-based neural network control of an upper-limb power-assist exoskeleton robot. In: Conference of proceedings of the 10th international conference on Advances in Neural Networks
7. Nathanael J (2011) Contributions à l'exploitation d'exosquelettes actifs pour la rééducation neuromatrice. Thesis prepared at the Institute of Intelligent Systems and Robotics (ISIR), pp 17–26
8. Ray LR, Stengel RF (1993) A monte carlo approach to the analysis of control system robustness. Automatica 29(1):229–236
9. Boeren F, Bruijnen D, Oomen T (2017) Enhancing feedforward controller tuning via instrumental variables: with application to nanopositioning. Int J Control Ident Control Nonlinear Electro-Mechanical Syst 90(4): 746–764

Part IV
Vibrations

Coupled Magnetic and CFD Modelling of a Structural Magnetorheological Vibration Absorber with Experimental Validation

M. Kemerli, T. Engin and Z. Parlak

Abstract Magnetorheological fluid is a smart material which can change its viscosity in milliseconds depending on the magnetic field applied. This brings a great advantage to create variable damping ability if it is used in an absorber. The stiffness of the absorber can be manipulated by an external magnetic field which effects the apparent viscosity of the magnetorheological fluid inside the absorber. Various control algorithms can be used to provide an effective absorption for any kind of structural vibration. Because of these features, magnetorheological absorbers have received great attention of researchers in the last decade. In this study, it is aimed to simulate a magnetorheological absorber under a sinusoidal vibration with Computational Fluid Dynamics and Magnetic Field Finite Elements Analysis. The magnetorheological fluid is modelled as a Non-Newtonian fluid and Herschel-Bulkley viscosity model is used to determine the apparent viscosity. Magnetic field is modelled for a constant current which generates different magnetic flux densities inside the absorber body. The Computational Fluid Dynamics and Finite Elements Analysis solutions are coupled in a two-dimensional axisymmetric domain and the results are revealed. The coupled solution of both are realized for the first time in the literature by means of an apparent viscosity approach. The numerical solution is compared with the experiments. A good agreement is observed between both results.

Keywords Magneto-Rheology · MR fluid · MR damper · Herschel-Bulkley CFD

1 Introduction

Magnetorheological (MR) fluids are a type of a smart material which can change its rheological properties in milliseconds depending on a magnetic field applied on. This feature brings a great advantage to use it in different type of absorbing

M. Kemerli (✉) · T. Engin · Z. Parlak
Sakarya University, Sakarya, Turkey
e-mail: mkemerli@sakarya.edu.tr

R. Rizk and M. Awad (eds.), *Mechanism, Machine, Robotics and Mechatronics Sciences*, Mechanisms and Machine Science 58,
https://doi.org/10.1007/978-3-319-89911-4_9

mechanisms. The MR fluid is used in vehicles as absorbers, clutches, brakes, and also used in aerospace, buildings, human prosthesis and so on [1, 2]. The essential feature of the fluid is that it can be controlled by an electrical current which generates a magnetic flux density and that changes the stiffness of the mechanism precisely. This also brings different disciplines together to design such mechanisms. Fluid mechanics and rheology in particular, structural mechanics, system dynamics and control, civil engineering, biology are affiliated for specific uses such as vibration control of buildings or human prosthesis. Even a chemistry knowledge is essential to manufacture an MR fluid. This makes the topic an interdisciplinary issue with a rich academic background. First explorations are goes back to 1940s [3, 4] and to this day thousands of academic work exist in literature for different types of usage of MR fluid.

MR absorbers are one of the hottest topic of MR fluid both in academy and industry which can be counted as the most fundamental use of MR fluid. The mechanism works with the same principle of conventional absorbers except the rheological properties of the fluid is manipulated by an external magnetic field so the stiffness of the absorber. The biggest advantage of it is its rapid response. A single MR absorber can work like an infinite number of conventional absorbers within its own range. From PID to neuro-fuzzy, hundreds of control algorithms are exist in the literature to control the MR damper depending on the problem. In general use, an embedded control algorithm determines the current and different type of sensors feedback the system. The strategy of control totally depends on the problem limitations, bounds or target.

The most important thing to control an MR damper is to know its characteristics and determine its response apart from the feedback values. That is why the researchers have been focused on a representative dynamic model of MR damper. There are lots of different dynamic models in the literature such as Bingham model, biviscous models, viscoelastic-plastic models, different types of Bouc-Wen models and so on [5, 6]. Unfortunately because of the hysteresis behavior of the MR damper it is very hard to model the exact response. Also in the design process the behavior of the damper can be considered as a closed box and it is not possible to predict the behavior with a dynamic model. This bring two different solutions to it. The first one is experimental method which is obviously not an easy way since it takes a lot of money and effort and the second one is to simulate the physics with a simulation tool which is a much better option for designers.

The CFD software and computer technology is way better than the past and it is developing in a pace. There are lots of models also generated for absorbers and it is possible to calculate generated force of a damper and uncover the underlying physics. For an MR absorber it is needed to link the magnetic flux density and viscosity of the fluid. Unfortunately most of the studies focused on control systems or dynamic models of MR dampers while there are limited works on the hydraulic structure and fluid domain of MR damper and CFD analysis in particular.

It is clear that a better link between the magnetic field and apparent viscosity can be formed. This study proposes a new multiphysic model with a 2D axisymmetric domain with less computational cost and with a better and coupled link between magnetic and fluid properties of fluid. The rheological behavior of the fluid is mod-

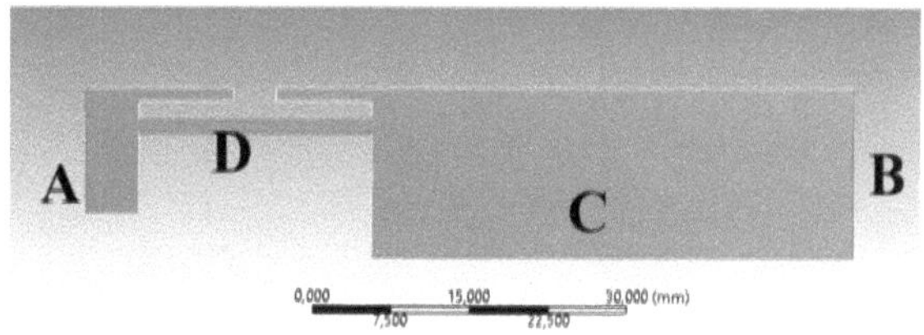

Fig. 1 Boundaries of the CFD model

elled by using an apparent viscosity approach which is quite simple and effective. Also this new model can reveal the underlying physics better. It is known that most of the models depend on assumptions with a considerable error while this model cut down the error with the experimental results.

This sections below consist of three different sections where in the second section the model is explained in all aspects, the third section shows and discusses the results while the final section concludes the study.

2 Overview of the CFD and Magnetic Model

To represent the physics of the problem there need to model two different physical phenomenon. The first one is the fluid dynamics and the second one is to model the magnetic effects on the fluid. To fulfil that the CFD analyses are conducted in ANSYS FLUENT software while the magnetic effects are simulated with ANSYS Magnetostatics tool. A 2D axisymmetric CFD model which is shown in Fig. 1 and a 3D axisymmetric Magnetic FEA model is created which shown in Fig. 3. In the CFD analysis, a coupled and an uncoupled solution is obtained to compare the coupled model.

2.1 CFD Model Description

The CFD model is created with four boundary conditions as shown in Fig. 1. A and B are the lower and upper boundaries of the absorber cylinder respectively. Instead of modelling a moving piston head, the A and B boundaries are moved and the center of the absorber model is kept stable to transfer the magnetic flux density data properly. C is the symmetry boundary to model the axisymmetry. The D region is the gap which the fluid is forced to pass through. The absorber creates the reactive force in that region with the viscous resistance and the magnetic field also increases the generating force by changing the apparent viscosity of the fluid.

The mesh structure is given in Fig. 2. A structured mesh is preferred and finer meshes are used inside the gap. The cells are almost perfectly quadratic and it is

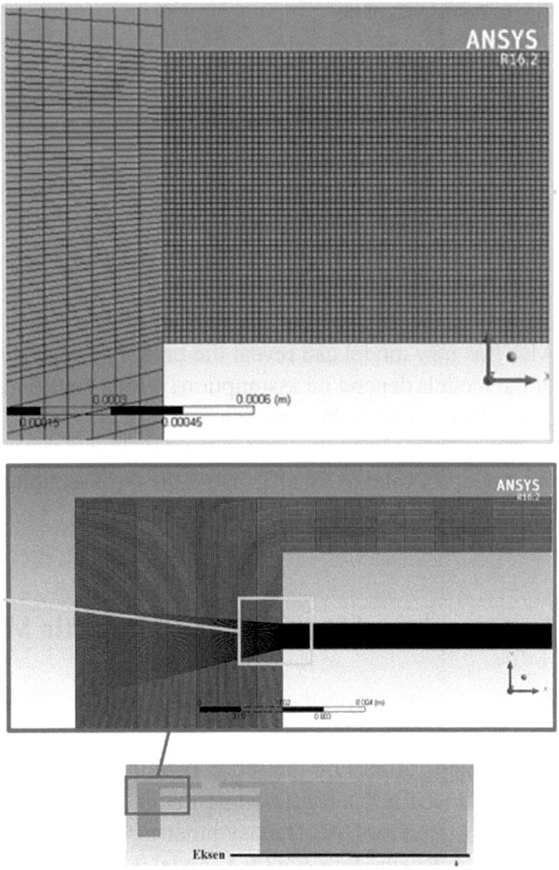

Fig. 2 Mesh structure of the domain

easier to transfer the magnetic data and create a dynamic mesh with such an ordered structure. 434460 cells and 438152 nodes are used. Using a 2D model gives a great advantage to create a successful mesh structure. The skewness is around 10E-4 and the orthogonal quality is 0.9995 in average which can be considered almost a perfect mesh structure.

2.2 *Magnetic Model*

To model the behavior of the MR fluid under different magnetic flux densities, the correlation between the magnetic field and magnetic flux density has to be known. The MRF132DG MR fluid is chosen for the model and the magnetic properties of the fluid which is published by the LORD Company is used [7].

Fig. 3 **a** The magnetic model of the domain **b** The magnetic flux vectors on the surface

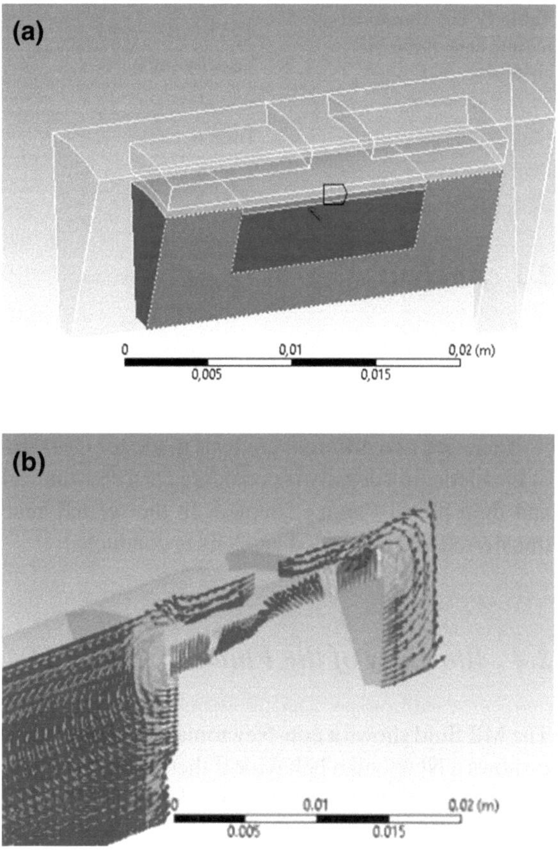

A numerical analysis is conducted with ANSYS Magnetostatics tool for a 3D absorber domain. The current is kept stable at 1 A to create a magnetic flux inside the fluid domain. The coil which generates the magnetic field shown in blue color in Fig. 3a with the damper head colored in grey. There are 220 turns exist for the coil and to minimize the computation time, 25° slice of the cylinder is chosen as the domain. The domain satisfies the conditions of an axisymmetric geometry and it is possible to transfer the surface data to the CFD domain. The magnetic flux vectors are shown in Fig. 3b which shows the development of the magnetic field in the entrance region of the gap. The color of the vectors represents the intensity of the magnetic flux density.

Table 1 The sinusoidal motion data of the MR absorber

Displacement (m)	$0.0075 \cos(\omega t)$
Velocity (m/s)	$-0.05 \sin(\omega t)$
ω (rad/s)	6.667
Time step (s)	0.001
Total time (s) $(2\pi/\omega)$	0.942

2.3 *Analysis*

A transient analysis is run for the fluid domain. The boundaries are making a sinusoidal motion to model the full cycle of a vibration and to compare it with the experimental data. The velocity and displacement values are shown in Table 1.

There are two different analysis made for CFD. In the first one, the average value of magnetic flux density is considered at a constant value in other words the magnetic and fluid physics aren't coupled. In the second analysis the magnetic field data is transferred and a coupled analysis is conducted.

2.4 *Rheology of the Fluid*

The MR fluid shows a non-Newtonian characteristics under any magnetic field while it shows a Newtonian behavior if there is no magnetic field applied on. To model the non-Newtonian behavior of the fluid there are lots of models exist in the literature for MR fluids [8]. The magnetic flux density effects the behavior of the fluid so the model has to be determined with the consideration of the magnetic flux density. There is a clear correlation between the magnetic flux density and the yield stress (τ_0) of the fluid.

The Herschel-Bulkley model is chosen to model the rheology of the fluid since it is one of the most common models to model the MR fluid [8]. The definition of the apparent viscosity for a Herschel-Bulkley model as follows [9];

$$\begin{cases} \mu = \frac{\tau_0}{\dot{\gamma}} + k\left(\frac{\dot{\gamma}}{\dot{\gamma}_c}\right)^{n-1} & |\dot{\gamma}| \geq \dot{\gamma}_c \\ \mu = \tau_0 \frac{\left(2-\frac{\dot{\gamma}}{\dot{\gamma}_c}\right)}{\dot{\gamma}_c} + k\left[(2-n) + (n-1)\frac{\dot{\gamma}}{\dot{\gamma}_c}\right] & |\dot{\gamma}| \leq \dot{\gamma}_c \end{cases}$$

μ stands for the apparent viscosity, $\dot{\gamma}$ is deformation rate, $\dot{\gamma}_c$ is the critical deformation rate, τ_0 is the yield stress. k and n values represents the consistency index and flow index respectively. k, n and $\dot{\gamma}_c$ are kept constant while the τ_0 value is kept at a constant value for the uncoupled model and defined depending on the magnetic flux density for the coupled model. The constants are given in Table 2.

Table 2 The rheological model constants for MR fluid

Density	3090 kg/m^3
k	0.092 Pa.s
n	1
τ_0 (for uncoupled model)	32000 Pa
$\dot{\gamma}_c$	900 s^{-1}

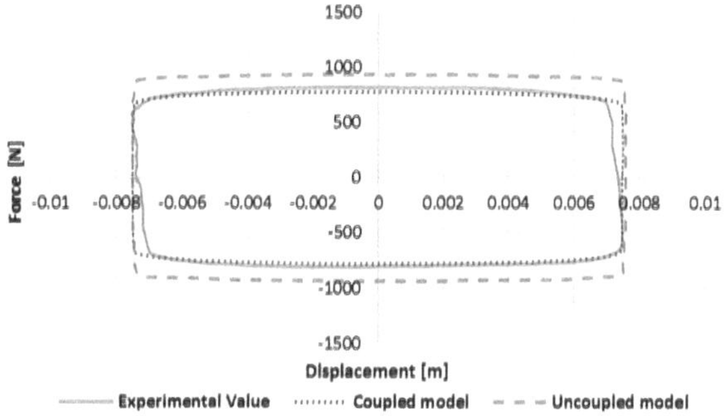

Fig. 4 Force-displacement chart for the absorber

The relation between the τ_0 and magnetic flux density is represented with a 4th order polynomial equation which derived from the published data of LORD Company by curve-fitting as follows [10];

$$\tau_0 = 51.962B^4 - 176.51B^3 + 158.79B^2 + 13.708B + 0.1442$$

To calculate the viscosity for each cell for the coupled model a User Defined Function (UDF) is written in C language and this modified equation is implemented for each iteration in the numerical analyses.

3 Results and Discussion

The results are compared with uncoupled solution and the experimental data which is obtained from previous works [11]. When we look up the force-displacement relation at Fig. 4 which is the fundamental output chart for an absorber; the maximum force value of 849 N is predicted for the coupled solution with a 3.5% error. The error is decreasing if the velocity gets smaller. If the results of the uncoupled model are evaluated, the values are far away from the experimental data and it is obvious that the coupling solution is amazingly successful when it is compared with the uncoupled one.

If the right and left bounds of the cycle are considered, the force is making a rapid fall because of the absorber changes its direction of velocity. In the experimental data, if looked carefully, the force has asymmetric values in both directions. In numerical

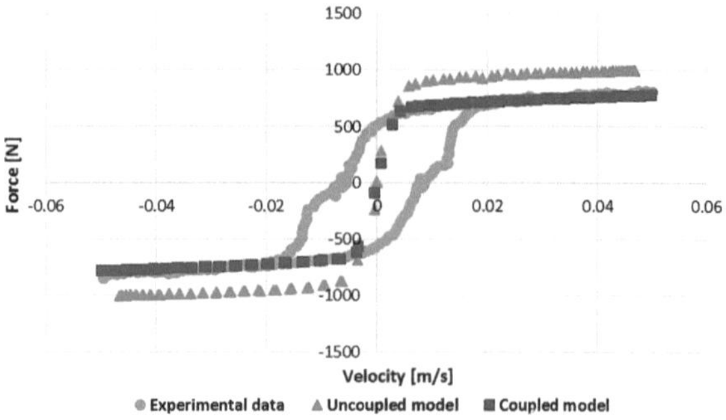

Fig. 5 Force-velocity chart for the absorber

analysis instead, a symmetric shape can be observed. This is because of the hysteresis characteristic of the absorber which cannot be modelled with CFD. But if we compare the predicted force along the motion which is more deserved to be considered as a design criteria, there is a perfect agreement between both results.

When the velocity-force relation investigated shown in Fig. 5, the coupled model has much better agreement. The result doesn't only satisfy the maximum speed ranges, it is also successful at the full range of the velocity except the hysteresis region because of the same reason mentioned above.

At Fig. 6 the viscosity values of coupled solution along the gap is shown. In (a) the fluid velocity is minimum and in contrast in (b) it is at its maximum. The values are colored in local scale for each time step. The effect of the magnetic flux density to the apparent viscosity can be seen clearly at the viscosity contours and the magnetic field effects at the outside of the gap region cannot be modelled with an uncoupled solution.

At the Fig. 7 the velocity profile for the coupled solution is shown in two different places in the gap. (a) shows the places where the fluid is subjected to the magnetic flux and (b) shows the places where there is no magnetic field on the fluid. The fluid has a laminar flow profile if there is no magnetic field applied while if it is subjected to the magnetic field it has a bulk flow profile. This is because the fluid has non-Newtonian characteristics and in the center of the flow, which is called the core region, the deformation rate doesn't reach the critical yield stress and moves like a bulk solid. This profile also shows a successful model of the fluid and reveals the underlying physics.

4 Conclusion

Modelling of MR absorbers with CFD is still an unclear issue for the engineers and it is not easy to choose an appropriate model. To model the entire physic is time consuming while making rough assumptions may end up with considerable error. It is shown that a 2D axisymmetric model can minimize the computational cost.

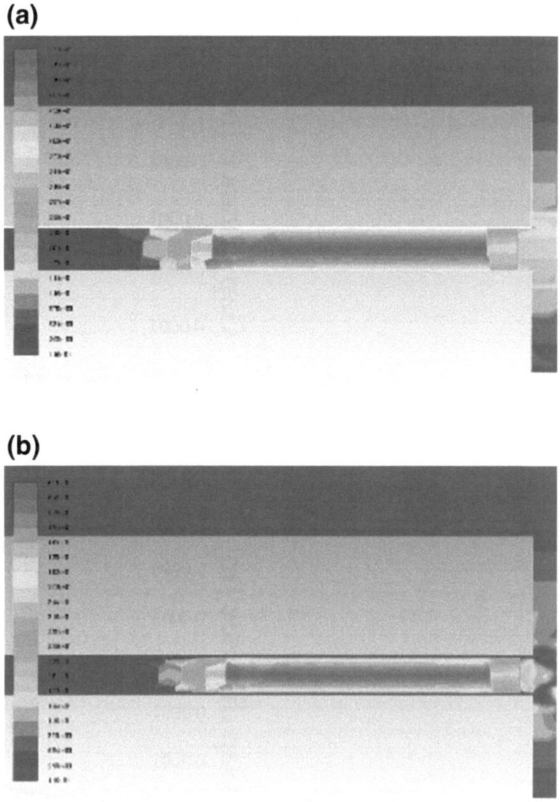

Fig. 6 Apparent viscosity contours **a** at minimum speed **b** at maximum speed

In the other hand, it is necessary to couple the magnetic effects for each cell. The study shows that using a classical CFD solution with several modification can show a perfect agreement with the experimental results.

For the magnetic effects it is enough to use a low-frequency analysis and there is no need to using a two way coupling model since the effect of fluid to magnetic properties seems negligible.

The study also shows that the hysteresis behavior of the fluid is still cannot be modelled with a CFD model which uses an apparent viscosity approach and there still need a better model to understand the underlying physics of the hysteresis behavior of MR fluid.

Acknowledgements The authors gratefully acknowledge TÜBİTAK for making this project possible under Grant No: 115M363. The authors acknowledge Sakarya University Scientific Research Projects Commission (SAÜ Bilimsel Araştırma Projeleri) for making this project possible.

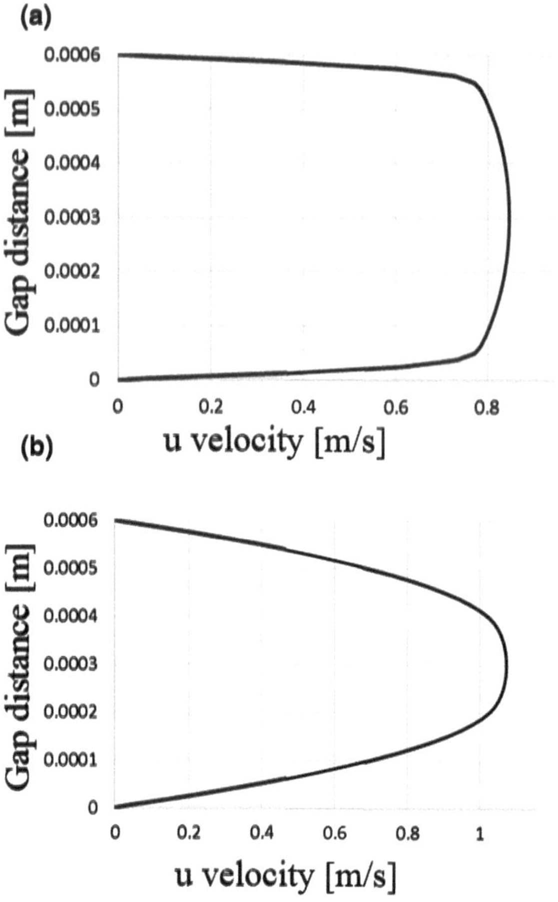

Fig. 7 Velocity profile **a** subjected to magnetic field **b** at zero magnetic field in the gap region

References

1. Guo S (2006) Dynamic modeling of magnetorheological damper behaviors. J Intell Mater Syst Struct 17:3–14
2. Baranwal D, Deshmukh TS (2012) MR-fluid technology and its application—a review. Int J Emerg Technol Adv Eng 2:563–569
3. Winslow WM (1949) Induced fibration of suspensions. J Appl Phys 20:1137–1140
4. Rabinow J (1948) The magnetic fluid clutch. Trans Am Inst Elect Eng 67:1308–1315
5. Wang DH, Liao WH (2011) Magnetorheological fluid dampers: a review of parametric modelling. Smart Mater Struct 20:23001
6. Spencer BFJ, Dyke SJ, Sain MK, Carlson JD (1997) Phenomenological model for magnetorheological dampers. J Eng Mech 123:230–238
7. Data T (2011) MRF-132DG magneto-rheological fluid. Lord Prod Sel Guide Lord Magnetorheol Fluids 54:11
8. Ghaffari A, Hashemabadi SH, Ashtiani M (2015) A review on the simulation and modeling of magnetorheological fluids. J Intell Mater Syst Struct 26:881–904
9. ANSYS (2013) ANSYS FLUENT user's guide. ANSYS FLUENT user's guide 15317:2498

10. Nguyen Q-H, Choi S-B, Wereley NM (2008) Optimal design of magnetorheological valves via a finite element method considering control energy and a time constant. Smart Mater Struct 17:25024
11. Parlak Z, Engin T (2012) Time-dependent CFD and quasi-static analysis of magnetorheological fluid dampers with experimental validation. Int J Mech Sci 64:22–31

Self-running Fault Diagnosis Method for Rolling Element Bearing

S. Kass, A. Raad and J. Antoni

Abstract While the machine is running, damaged components of the bearing trigger vibrations in the structure of the machine when it contacts other surfaces. These components appear at specific frequencies dictated by the geometry of the bearing and its rotation frequency. An autonomous fault detection method is therefore needed to improve the performance and the reliability of the mechanical system. This article aims to present an autonomous bearing fault detection process. This process takes into account the slip phenomenon by calculating a normalized indicator related to the existence of a bearing fault in a narrow band centered at the theoretical frequency. The latter is calculated from the geometry of the bearing, after preprocessing steps in order to equalize the baseline spectrum and to set an appropriate statistical threshold. An application on real data from the IMS database will be held at the end in order to detect and classify mechanical faults.

Keywords Autonomous fault detection · Slip phenomenon · Pre-processing Threshold · Detection and classification

1 Introduction

Rolling element bearing (**REB**) is one of the crucial elements in rotating machinery, and its failure is one of the most common problems that occur with a high probability [1]. Therefore, its corresponding faults (ball, cage, inner and outer race fault) should

S. Kass (✉)
Lebanese University, Beirut, Lebanon
e-mail: Souhaybkasss@gmail.com

S. Kass · J. Antoni
INSA Lyon, Lyon, France

A. Raad
Engineering Faculty, Lebanese University, Tripoli, Lebanon

J. Antoni
University of Lyon, Lyon, France

R. Rizk and M. Awad (eds.), *Mechanism, Machine, Robotics and Mechatronics Sciences*, Mechanisms and Machine Science 58,
https://doi.org/10.1007/978-3-319-89911-4_10

be diagnosed at an early stage. Those faults appear at characteristic frequencies that could be calculated based on the dimensions of the bearing, without taking into consideration the slip phenomenon. Such unavoidable phenomenon implies a small deviation of the characteristic fault frequencies from the expected ones [2].

In general, corresponding vibrations are generated when the subsequent contact between the damaged components and other surfaces of the bearing belongs to a second order cyclostationary (CS2) family [3]. This family is characterized by a periodic auto-covariance function. Several methods have been developed to extract spectral information using cyclic analysis, such as the squared envelope spectrum (SES), the spectral correlation (SC) and its normalized version called the spectral coherence (SCoh) [4–7]. These methods give us spectral indicators, where the presence of peaks at characteristic cyclic frequencies could be an indicator of a specific fault. In practice, this information is usually hidden by noise due to the complexity of transmission mechanisms and the surrounding noisy vibrations. The latter create components that are unrelated to the faults in the bearing, but to other sources leading to a noisy spectrum. The harmonics related to the shaft speed can be eliminated by synchronous averaging techniques [8] or by comb filtering [5].

Based on foregoing, the construction of an autonomous algorithm must take into account the following points:

1. the most efficient and robust frequency representation method; since each fault is characterized by its own frequency,
2. the slip phenomenon involving a deviation from the expected fault frequencies,
3. the proposition of an algorithm similar to the visual human inspection,
4. the time and computational complexity.

This paper is organized as follows. Section 2 presents a summary of different advanced frequency representations in the context of the bearing diagnosis. In Sect. 3, a new method based on the frequency representation is introduced and applied to an illustrative real example. Section 4 includes applications and results of the autonomous method and its response to fault severity. Conclusions and perspective will be drawn at the end of this article.

2 Context on REB Diagnosis

2.1 *Review on Spectral Analysis*

Envelope analysis (ES) has been admitted as a reference technique for the REB diagnosis under constant speed [9]. Typically, it consists of a filtering step around a high-frequency band in which the impulsive excitation is amplified before demodulating the obtained signal to form the envelope. The spectrum of the latter includes the desired information [5].

Some techniques based on the kurtogram [10] are developed to find the best frequency band. Later, the Hilbert transform of the obtained signal is used to calculate the ES. It was shown in [11] that it is preferable to use the SES instead of ES inasmuch as the latter introduces foreign components that appear as misleading peaks in the spectrum.

A more powerful alternative, the SC and its normalized version SCoh, can distinguish the modulation effects due to rotation of the gears and rotations of the internal bearing ring [4]. Despite its advantages over the ES, its use has been hampered by its high computational cost. A fast estimator is the Cyclic Modulation Spectrum (CMS) [12]. It is the Fourier transform of the spectrogram; therefore, it suffers from the uncertainty principle and has some limitations to detect relatively slow periodic modulations. Recently, [6] corrected that situation by proposing a fast estimator; the Fast-SC ($S_x^{Fast}(\alpha, f)$) which is based on the short-time Fourier transform (STFT).

As advocated in [13], the SCoh is used instead of SC. It is interpreted as the SC of the whitened signal, which tends to equalize regions with very different energy levels and thus to magnify weak cyclostationary signals. The SCoh is calculated as follows:

$$\boldsymbol{\Upsilon}_x^{Fast}(\alpha, f) = \frac{S_x^{Fast}(\alpha, f)}{[S_x(f + (\alpha/2) . S_x(f - (\alpha/2))]^{1/2}} \tag{1}$$

where f is the spectral frequency, α is the cyclic frequency and $S_x(f)$ is the power spectral density (PSD) of the raw time signal x().

The integration of Υ_x^{Fast} over f gives an improved version of the SES. There are two ways to calculate this integration [6]:

$$s_x^{EES}(\alpha) = \int_{f1}^{f2} |\Upsilon_x(f, \alpha)| \, df \tag{2}$$

$$s_x^{SES}(\alpha) = \left|\int_{f1}^{f2} \Upsilon_x(f, \alpha) \, df\right| \tag{3}$$

The first one is expected to enhance better the non-zero cyclic components than the SES. SES integrates complex values, which may converge towards zero in some cases of fast rotating phases.

2.2 Bearing Fault Signature

The expected patterns in the EES for best-known faults of the REB (ball, cage, inner and outer ring), varies with the variation of fault types and its theoretical frequencies are calculated from the following formulas:

Ball Pass Frequency Outer race (BPFO)

$$\text{BPFO} = \frac{nf_r}{2}\left\{1 - \frac{d}{D}\cos\varphi\right\} \tag{4}$$

Ball Pass Frequency Inner race (BPFI)

$$BPFI = \frac{nf_r}{2}\left\{1 + \frac{d}{D}\cos\varphi\right\} \tag{5}$$

Ball Pass Frequency Inner race (BPFI)

$$FTF = \frac{f_r}{2}\left\{1 - \frac{d}{D}\cos\varphi\right\} \tag{6}$$

Ball Spin Frequency (BSF)

$$BSF = \frac{D}{2d}\left\{1 - (\frac{d}{D}\cos\varphi)^2\right\} \tag{7}$$

where f_r is the shaft speed, n the number of rolling elements, φ is the contact angle, d the REB diameter and D the pitch circle diameter.

These frequencies are calculated from the geometry of the bearing assuming a perfect rolling motion of the rolling element. Whereas in reality, a random frequency deviation is involved (usually at the order of 1–2% from calculated ones [2, 5, 14]) and are located in the vicinity of the theoretical frequencies due to the slip phenomenon. In details, the angle contact varies with the position of each rolling element, since the ratio of the radial load to the evolution of the axial load changes, giving birth to this phenomenon. Thereby, each rolling element will have a different diameter and will try to roll at different speed, while the cage limits the deviation of the rolling elements from their mean position, thus causing some random slip [5]. This random deviation, despite being small, compromises the fault follow-up on a frequency contrary to the gear cases. It implies by consequence the need to look for the fault in a centered narrow band having a width of 1–2% as mentioned above.

An outer race fault involved in a constant modulation is characterized only by harmonics of the BPFO. Nevertheless, inner race and ball fault involving a periodic amplitude modulation is characterized by harmonic of the BPFI (respectively BSF), and indeed of the sub-bands equal to *fr (respectively FTF)*. Therefore, in order to track the signature of BPFI (respectively BSF), the research must be extended to the sub-bands as can be seen in Fig. 1.

3 Proposed Method

The following method intends to simulate the human visual inspection to differentiate significant peaks from background noise. Two thresholds are introduced for this purpose. The first one is calculated from the histogram in order to determine whether the magnitude carries meaningful information or not. The second one—which rep-

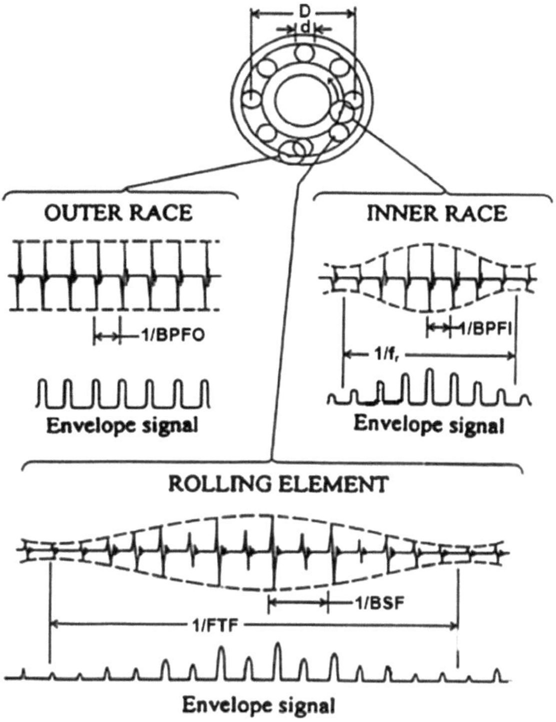

Fig. 1 Typical signals and envelope signals from local fault in REB [4]

resents the median of the signified peaks- is used to determine if this magnitude is fairly significant compared to other magnitudes along the frequency axis.

Once the ES is calculated, the threshold needs to be set. It is necessary that the ES baseline be equalized against background noise (Fig. 2). For this purpose, a pre-processing step is requested and particularly in the case when the Hilbert based-SES is used or the nonlinear filtering of the SCoh cannot be applied.

3.1 Pre-processing Step-Baseline Equalization

The aim of this part is to equalize the signal with respect to background noise so that the baseline is constant along the frequency axis in order to construct a statistical threshold $\lambda_{1-p}(\alpha)$ where p is the risk probability [15]. This threshold is based on the histogram of the (ES) in order to model a probability distribution [16]. Then, it is used to determine whether peaks are characteristics of the fault or not.

To be robust, the baseline equalization should not destroy significant information linked to bearing fault. So a rectangular window w_N of size N points—of which N is a variable adapted to each signal—is slipped over the frequency spectrum, in which the median is calculated, which gives SES_{Med}, represented by the bottom red

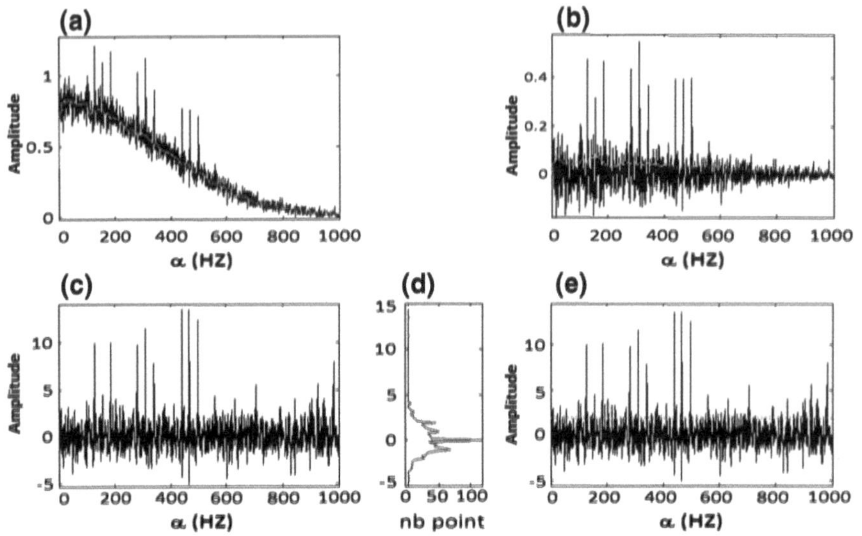

Fig. 2 **a** SES of vibration signal and its sliding median in dashed red line **b** centred spectrum after removing sliding median and its MAD in dashed red line **c** equalized ES after dividing by MAD **d** Histogram of **c**, **e** threshold λ_{1-p} (f) on equalized spectrum

line—the signature of background noise- (Fig. 2a). Subtraction of SES_{Med} cannot affect the outliers and gives a zero centered spectrum without equal level (Fig. 2b).

Another necessary step is to divide the latter spectrum (SES − SES_M) by the median absolute deviation MAD of itself calculated using a sliding window (Fig. 2c).

Once the spectrum is set at an equal level, the statistical threshold λ_{1-p} (α) is calculated based on the probability density function (PDF).

3.2 Proposed Algorithm

Once the equalized ES is obtained and the thresholds are selected, the autonomous algorithm is applied, and its steps are detailed as follows:

Step 1: Define the band centered at each frequency α_{c1} and its harmonics α_{cl}, with a deviation that varies 2% around the first band and 1% around the others due to the iterative correction of the central frequency of the next bands. The first deviation is used in order to respect the deviation caused by the slip phenomenon. Despite the uniform response behavior of the harmonics for the cited phenomenon, and in order to avoid the deviation that is associated to frequency resolution, the 1% narrow band is used.

$$\alpha_{c1} \le \frac{\alpha_{max}}{3} \tag{8}$$

where α_{max} is the maximum cyclic frequency and 3 is the number of tracking harmonics.

Step 2: In each narrow band, identify the characteristic frequency of the actual damage, where the highest peak is found in the following way:

$$m_{bl} = max\left(EES_{Band_i}(\alpha)\right) \tag{9}$$

where $Band_i$ is the band around the harmonic i.

Step 3: Once the m_{bl} is identified, if its value exceeds the fixed threshold $\lambda_{1-p}(\alpha)$, a score p_l based on the relative error between m_{bl} and the fixed threshold, will be calculated according to the following formula:

$$p_l = \frac{m_{bl} - \lambda_{1-p}(\alpha)}{m_{bl}} \tag{10}$$

In addition, this step includes a center correction of the next band based on the actual frequency of the peak, to avoid band enlargement and consequently foreign peaks:

$$\alpha_{e_{(i+1)}} = \frac{i+1}{i}\sum_{n=1}^{i}\alpha_{e_n} \tag{11}$$

where α_{e_n} is the central frequency of the band number n and α_{el} is the actual frequency.

Step 4: The total score associated to the fault existence is equal the average of all calculated scores:

$$p_{def} = \frac{1}{nb\ band}\sum_{l=1}^{nb\ band} p_l \tag{12}$$

$$p_{right} = \frac{1}{\text{nb band}}\sum_{l=1}^{nb\ band} p_{lr} \tag{13}$$

$$p_{left} = \frac{1}{nb\ band}\sum_{l=1}^{nb\ band} p_{ll} \tag{14}$$

p_{def}, p_{right}, p_{left} are normalized scalar values associated with fault existence and its possible left and right side band, that differ from the other analyses types which require an expert in the diagnosis field to process data. Nevertheless, those values can take any value between zero and one. So it is necessary to compare it with a threshold in order to give a significant conclusion.

This step also includes another sub-step, only if $\mathrm{p_{def}} > T$, in the case of searching the sub-band existence. In the same way, the left and right scores (p_{lr}, p_{ll}) of the possible side bands are calculated for each harmonic by scanning the spectral frequency around the fault frequency and its harmonics, Usually, the wide band up to 20% around the fault frequency is considered.

The reason to calculate two different scores for the left and the right sub-bands is that the two sub-bands are not necessarily symmetrical, and the existence of the

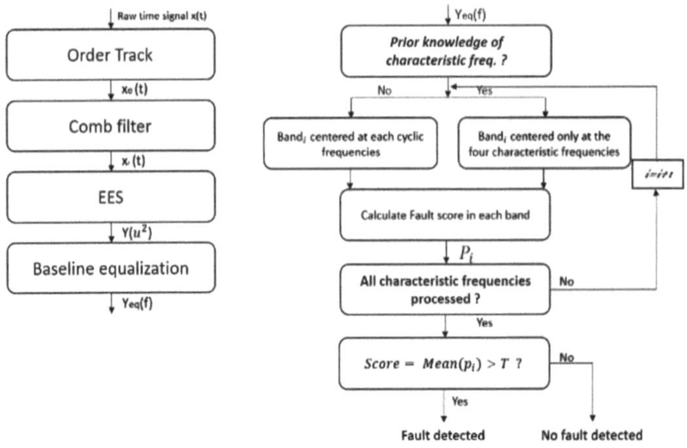

Fig. 3 Algorithm flowchart

one does not mean the existence of the other. For that, the combination of these two values can destroy the information.

Step 5: The detection threshold T is the relative error of the mean of all peaks exceeding the first threshold $\lambda_{\text{detection}}$ along the cyclic frequency axis and the threshold. They are given by the following equations:

$$\lambda_{\text{detection}} = \frac{1}{\text{Nbpic}}\left[\sum \text{SES}(\alpha) : \text{SES}(\alpha) > \lambda_{1-\text{p}}(\alpha) \,\text{and}\, \alpha\, \text{bi}\right] \tag{15}$$

$$T = \frac{\lambda_{detection} - \lambda_{1-p}\,(\alpha)}{\lambda_{detection}} \tag{16}$$

where $Nbpic$ is the number of points exceeding $\lambda_{1-\text{p}}\,(\alpha)$ and bi the set of narrow band i.

If p_{def} has a higher value than T, the peaks are quite aberrant with respect to the noise and other peaks along the frequency axis, hence the sought fault exists. In fact, the fault decreases with the increase in number of harmonics.

The complete flowchart for the algorithm described in this section is shown in Fig. 3.

If a prior knowledge on bearing characteristics is provided, the four characteristic frequencies are calculated. Then, a narrow band is centered on each characteristic frequency and the score P_i specify if one of the faults is presented. If no prior knowledge is given, a narrow band is selected and the score P_i is calculated for all frequencies to check the presence of faults.

In the case, where the characteristic frequencies are known, and in order to reduce the calculation time, small modifications are introduced to the above code in step 1 and step 4 as follows:

Step 1: define the search band centered only to the four theoretical damage characteristic frequency α_{c1}, and its harmonics see Fig. 5.

Step 4: in case of BPFI or BSF, the sub-band search is reduced to only the *fr* and *FTF* respectively. (see Fig. 7).

3.3 Algorithm Illustration

To illustrate this algorithm, a simple case is used wherein the four characteristic fault frequencies are already known. A real faulty bearing signal is used. The algorithm has been applied to a raw time signal shown in Fig. 3. This signal is a case where the bearing has an inner race faults coming from (CRWU) Bearing Data Center (illustrated in Sect. 4). According to bearing's geometry given by the provider of this database, the characteristic frequencies of the faults are calculated as a function of rotation frequency, and the $BPFI = 5.4152 \times fr$ (Hz), where fr is the shaft speed.

The Fast-SC-based EES is calculated according to formula (8). For all the other sections, the window length in the Fast-SC is set to N_w (equal to 2^7 here) in order to achieve a frequency resolution Δf *of* 100 Hz. The cyclic range α_{max} is chosen so as to include at least three harmonics of the highest fault frequency, viz the (BPFI) since the number of harmonics provides an indication of the fault severity. The EES is computed over the frequency range $[0{:}Fs/2]$ and hence referred to as wideband envelope spectra.

The EES of the vibration signal is represented in Fig. 5, and used as a reference example for this part.

The test bearing has an inner-race fault which is characterized by harmonics at BPFI (162 Hz) in the ES, and sub-bands equal to the shaft frequency (29.2 Hz).

Once the threshold (dashed line in Fig. 5) is calculated $\lambda_{1-p}(\alpha)$ equal to three times the standard deviation of a Gaussian distribution where $p = 2.5\%$. The fault existence is tested using the algorithm.

Figure 4 shows a raw time signal that has an inner race fault. A periodic amplitude modulation at rotation frequency is clearly seen at the zoomed signal. Figure 5 shows the corresponding frequency representation in which peaks at the BPFI and its harmonics appear. Therefore, a side-band appears at the *fr*. Figure 6 shows the way in which the algorithm tracks the BPFI and identifies the maximum in the drawn band (Step 1 and 2). While Fig. 7 shows the extended research to the BPFI side-bands (Step 4). Finally, the BPFI calculated score is equal to 0.9, p_{right} is equal to 0.87 and p_{left} is equal to 0.8. While T is equal to 0.5 and therefore the BPFI and its side bands are detected.

4 Application and Results

In order to validate the proposed algorithm, a benchmark REB is used, provided by NASA, from the Intelligent Maintenance Systems (IMS), University of Cincinnati.

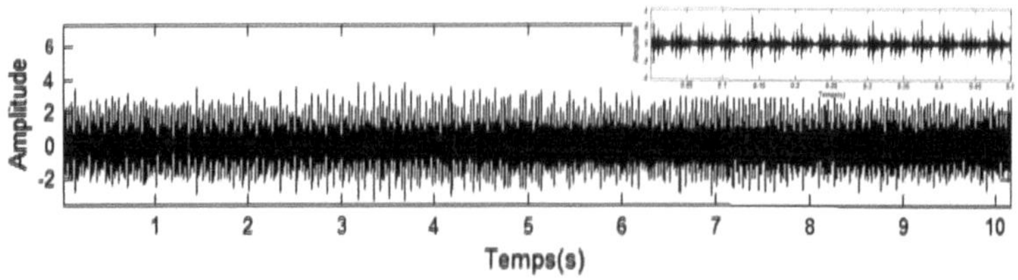

Fig. 4 Raw time signal of record 211DE, and its zoomed portion

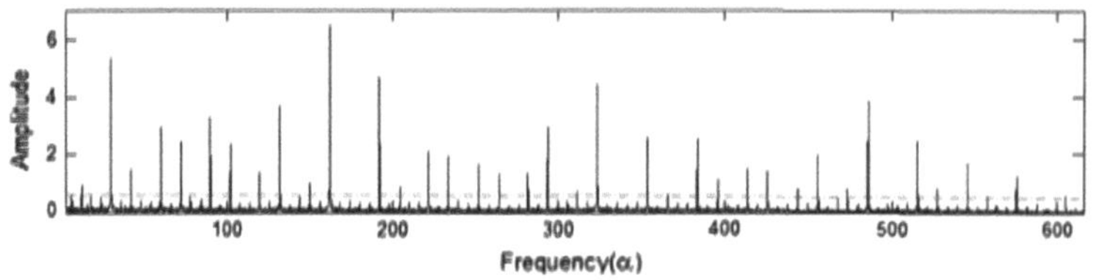

Fig. 5 EES of the raw signal, the red line represents the histogram based threshold

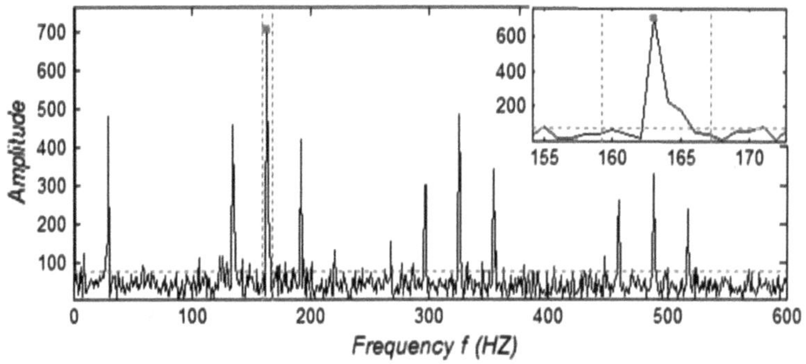

Fig. 6 Maximum identified in narrow band centered at BPFI

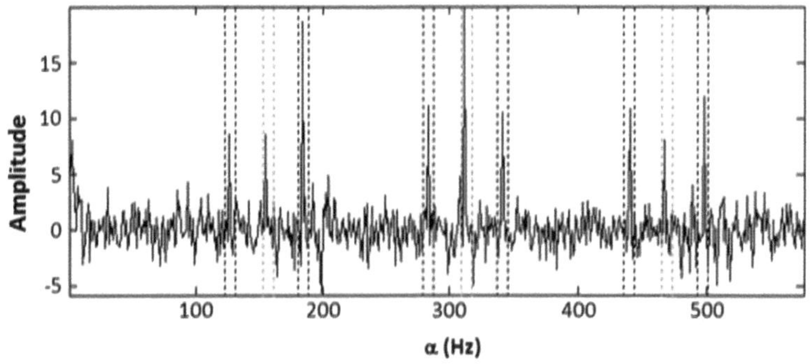

Fig. 7 Score research for: **a** fault frequency and its harmonics (dashed red line), the sub-b and at *fr* (dashed black lines)

Table 1 Characteristic frequencies of the test rig

Rotation frequency (*fr*)	33.3
BPFO	236
BPFI	297
BSF	278
FTF	15

This database has become a reference used to test new algorithms in order to compare their efficiency against existing techniques.

In order to show how this autonomous method deals with the fault evolution process, the run-to-failure bearing test database is used. This test is performed at a constant speed of 2000 rpm dictated by an AC motor on a test rig under normal load conditions. The bearing test rig hosts four identical bearings of type Rexord ZA-2115 in the same shaft. Based on its characteristics (16 rollers, pitch diameter of 7.15 cm, roller diameter of 0.84 cm and contact angle of 15.17°) and the equations given in section B. Table 1 shows the four fault frequencies.

Table 2 contains all related information about the used dataset from the IMS. A detailed explanation about this database has been given in [17].

Assuming that no information about the fault type is known, the above-detailed method will be applied in order to look for the existence of the main rolling faults presented. Figure 8, represents the ES of the whole database which includes 984 signals from the run (beginning) to the failure (end). The four scores associated with each bearing fault are represented in Fig. 8b. Figure 9, represents the same results in each accelerometer. The results presented in Figs. 8 and 9 show that only the BPFO is detected in all accelerometers but at different instants. The change is related to the number of peaks found and their magnitudes which are dictated by the fault severity that may explain the drowned path. The latter is divided into two stages, the first one till 3.4 days where no fault signature is detected. The second stage begins hence with a significant increase of the score above the detection threshold. This threshold is nothing more than a CS2 indicator in the whole spectrum. The fluctuating trend can be explained by the damage evolution process. When the surface fault initiated, small cracks were formed and were later smoothed by the continuous rolling contact. As the damage spread over a broader area, the vibration level rises again.

For the first accelerometer, the BPFO began after 3.4 days while for other accelerometers, this time detection varies due to the change in position of the

Table 2 Dataset description

	Number of files	Number of channels	Endurance duration	Duration of recorded signal	Announced damages at the end of the endurance
Dataset 2	984	4	9840 min 6 days 20 h	16 min	Bearing 1: outer race

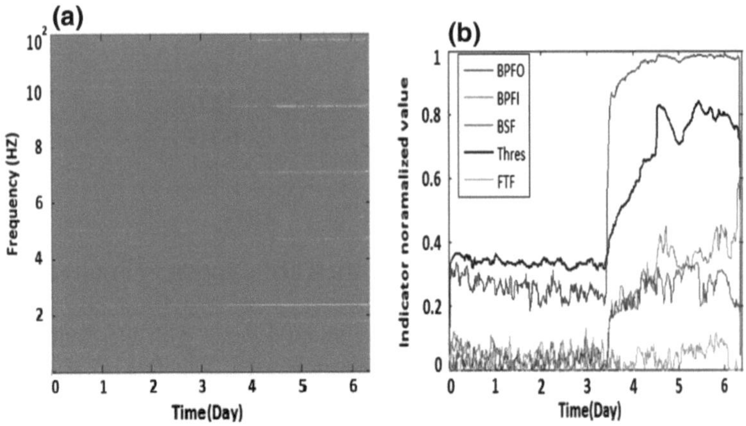

Fig. 8 Result of algorithm appliance on Dataset 2: **a** EES map **b** indicator values

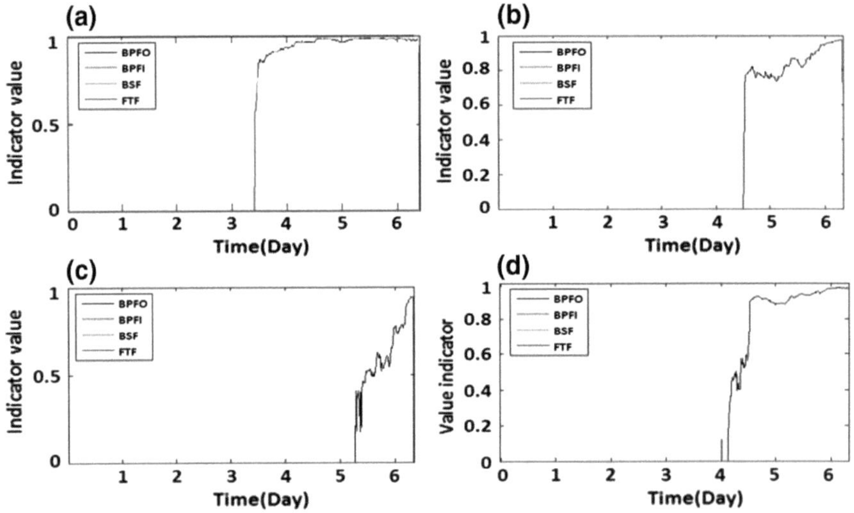

Fig. 9 Score value of accelerometers: **a** one **b** two **c** three **d** four

accelerometers with respect to the faulty bearing, as well as with the change in load direction. This method allows not only to detect and localize the existing fault types, but also the faulty bearing. In Fig. 9, the fault signature is detected in accelerometers 2 at 4.5 days, for accelerometers 3 at 5.3 days, while the accelerometers 4 where the radial load is in the same direction as the accelerometers 1. The fault is detected at 4.3 days preceding both accelerometers 2 and accelerometers 3.

In conclusion, the goal is reached and the indicator successfully matched the visual interpretation extracted from the ES map and reflected the fault severity. As a summary, it is quoted in Table 3.

Table 3 Summary of dataset 2 results

	Outer race fault			
	acc1	acc2	acc3	acc4
Results	TP (3.4 days)	TP2 (4.5 days)	TP2 (5.3 days)	TP2 (4.1 days)

5 Conclusion

Frequency representation, slip phenomenon, spectrum's baseline, threshold imposing and computational costs are factors that affect the autonomous bearing diagnosis. All these factors are taken into account in the proposed method. First, the Fast-SC EES was calculated from the spectral coherence. Then, an equalization of the EES baseline was done, and the threshold was set. The deviation caused by the slip phenomenon was encountered by searching the characteristic fault frequency into a narrow band rather than at a specific frequency. In each narrow band, a score related to fault existence was calculated. Finally, the total score, which is the mean of all calculated scores, was compared to the threshold and results were obtained.

The proposed algorithm has been explained, detailed and tested using real signals from IMS database in order to show its response to fault existence and development. As a result, the proposed method can easily detect the most frequent bearing faults and reflect their severity. Therefore, the aim, which was to implement the visual human inspection, was successfully achieved.

References

1. Renforth L, Hamer P, Clark D, Goodfellow S, Tower R (2013) Continuous, remote on-line partial discharge (OLPD) monitoring of HV EX/ATEX motors in the oil and gas industry. In: Petroleum and chemical industry technical conference (PCIC), 2013 record of conference papers industry applications society 60th annual IEEE, 2013
2. Pennacchi P, Borghesani P, Ricci R, Chatterton S (2011) An experimental based assessment of the deviation of the bearing characteristic frequencies. In: 6th international conference acoustic and vibratory surveillance methods and diagnostic techniques, Compiegne, 2011
3. Antoni J (2009) Cyclostationarity by examples. Mech Syst Signal Process 23:987–1036
4. Randall RB, Antoni J, Chobsaard S (2001) The relationship between spectral correlation and envelope analysis in the diagnostics of bearing faults and other cyclostationary machine signals. Mech Syst Signal Process 15:945–962
5. Randall RB, Antoni J (2011) Rolling element bearing diagnostics—a tutorial. Mech Syst Signal Process 25:485–520
6. Antoni J, Xin G, Hamzaoui N (2017) Fast computation of the spectral correlation. Mech Syst Signal Process 92:248–277
7. Gardner W (1986) Measurement of spectral correlation. IEEE Trans Acoust Speech Signal Process 34:1111–1123
8. Braun S (2011) The synchronous (time domain) average revisited. Mech Syst Signal Process 25:1087–1102
9. McFadden P, Smith J (1984) Model for the vibration produced by a single point defect in a rolling element bearing. J Sound Vib 96:69–82

10. Antoni J (2007) Fast computation of the kurtogram for the detection of transient faults. Mech Syst Signal Process 21:108–124
11. Ho D, Randall RB (2000) Optimisation of bearing diagnostic techniques using simulated and actual bearing fault signals. Mech Syst Signal Process 14:763–788
12. Antoni J, Hanson D (2012) Detection of surface ships from interception of cyclostationary signature with the cyclic modulation coherence. IEEE J Ocean Eng 37:478–493
13. Antoni J (2007) Cyclic spectral analysis of rolling-element bearing signals: facts and fictions. J Sound Vib 304:497–529
14. Smith WA, Randall RB (2015) Rolling element bearing diagnostics using the Case Western Reserve University data: a benchmark study. Mech Syst Signal Process 64:100–131
15. Pimentel MA, Clifton DA, Clifton L, Tarassenko L (2014) A review of novelty detection. Signal Process 99:215–249
16. Brandt A (2011) Noise and vibration analysis: signal analysis and experimental procedures. Wiley, p. 85
17. Qiu H, Lee J, Lin J, Yu G (2006) Wavelet filter-based weak signature detection method and its application on rolling element bearing prognostics. J Sound Vib 289:1066–1090

Effects of Yoke Thickness Change on the Vibration Behavior of an Electric Machine

H. Issa, N. Khishen, F. Druesne, J. Hallal, M. Hammoud and A. Hallal

Abstract The aim of this paper is to present a sensitivity study of the effect of the geometrical changes on the vibration behavior of an electric machine at the stator level. A wound rotor synchronous machine is chosen here. A finite element model (FEM) of tetrahedral elements of the stator is created using ANSYS-APDL. This paper, based on the changed geometrical parameter, presents 2 studies: Variation of the stator yoke thickness and the variation of the stator yoke thickness while keeping the same mass. In both studies, the decrease in the yoke thickness leads to the decrease in the natural frequencies although the mass remained constant. The decrease in the yoke thickness increases the stator flexibility which means that the stiffness is more effective than the mass.

Keywords Vibration · Finite element · Modal analysis · Electric motor
ANSYS APDL · Sensitivity study

H. Issa · N. Khishen · J. Hallal (✉) · M. Hammoud · A. Hallal
International University of Beirut, Beirut, Lebanon
e-mail: jaafar.hallal@liu.edu.lb

H. Issa
e-mail: 31130492@students.liu.edu.lb

N. Khishen
e-mail: 71130233@students.liu.edu.lb

M. Hammoud
e-mail: mohamad.hammoud@liu.edu.lb

A. Hallal
e-mail: ali.hallal@liu.edu.lb

F. Druesne
Laboratoire Roberval, UMR 7337, Sorbonne Universités, Université de Technologie, Compiègne, France
e-mail: frederic.druesne@utc.fr

R. Rizk and M. Awad (eds.), *Mechanism, Machine, Robotics and Mechatronics Sciences*, Mechanisms and Machine Science 58,
https://doi.org/10.1007/978-3-319-89911-4_11

1 Introduction

In the automotive applications, the noise of the electric machines is an important problem [1]. In this context, several studies focus on the vibro-acoustic modeling of the electric machines. These models will be used to predict and optimize these machines [2]. However, the results depend on the accuracy of the model.

One of the critical issues that affect the accuracy of finite element models in this domain is the material and the geometrical variability [3]. Actually, Boubaker et al. show a high noise level variability (10–20 dB) for 10 nominally identical vehicles [4].

This paper focuses on the sensitivity study of the yoke thickness with or without mass change of the mass of the stator. The yoke thickness takes several values and the resultant natural frequencies will be compared to the nominal natural frequencies in order to understand the effect of the yoke thickness on the vibration behavior.

Since we are working in a vibro-acostic context, only frequencies between 20 Hz and 20 kHz will be taken into consideration.

2 Finite Element Model

The machine chosen for this study is a wound rotor synchronous machine where the nominal dimensions are presented in Table 1.

First, a finite element model of the stator is created used ANSYS-APDL. This model is then subjected to a mesh convergence study.

2.1 *FEM Creation*

ANSYS APDL is the finite element software chosen to work on due to its flexibility during work. A FEM model can be created using a code text file in APDL. This feature is very important since variability or simply sensitivity study is targeted. All results are saved automatically for each simulation in a separated text file. This FEM model presented in Fig. 1 is suitable for a sensitivity study of any geometrical or material

Table 1 Nominal dimensions of the stator

Stator inner diameter	54 mm
Stator outer radii	100 mm
Stator middle radii	71 mm
Stator teeth width	5.5 mm
Stator slot height	17 mm

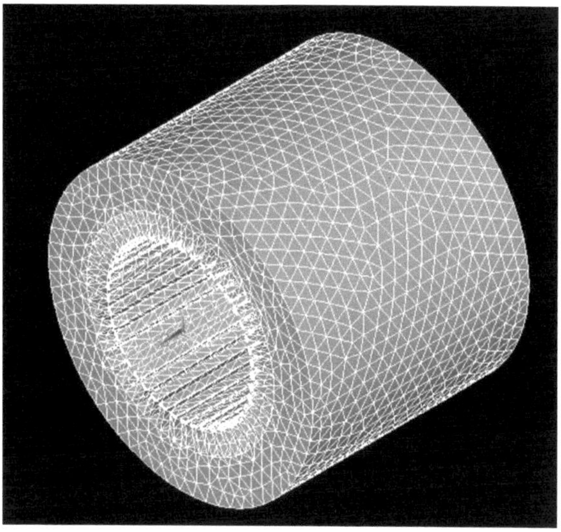

Fig. 1 Meshed stator model

parameters. Actually, the dimensions and the material properties are modeled as variables in the code text file.

2.2 *Mesh Convergence*

A mesh convergence study was done on the finite element model in order to well choose the mesh size. A compromising between the calculation time and the accuracy of the results is needed. Small element size (high number of nodes) leads to an accurate result. However, big element size (low number of nodes) leads to a rapid calculation [5].

Seven different mesh sizes, varying from 6000 to 50000 nodes were tested, and the results of 4 arbitrary modes are presented in Table 2.

Table 2 Number of nodes for different meshes

Mesh	Number of nodes
M1	6188
M2	9222
M3	9848
M4	13627
M5	16129
M6	19896
M7	50381

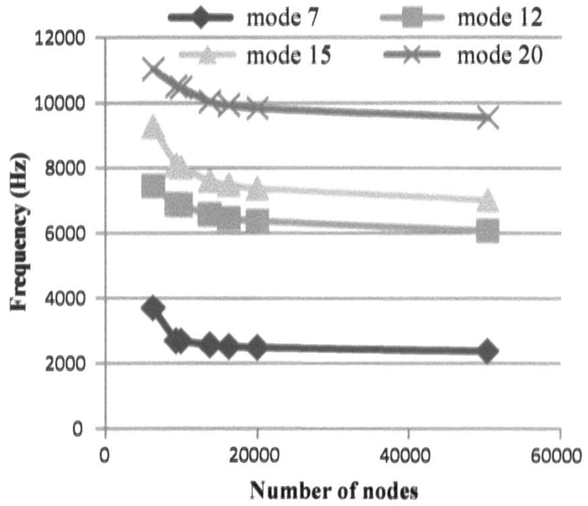

Fig. 2 Natural frequencies (Hz) of modes 7, 12, 15, 20 with respect to different meshes

Figure 2 shows the natural frequencies associated to 4 arbitrary mode shapes calculated on different mesh sizes.

With a convergence error, about 2.5% between meshes M4 and M7, it is clear that mesh M4 meets the convergence criterion.

In the following, mesh 4 with 13627 nodes will be used.

3 Results

Since the aim of this paper is on the effect of the yoke thickness, two sensitivity studies will be presented. The first one presents the change in the yoke thickness with a variable mass of the stator. While in the second one, the yoke thickness is variable with a fixed mass of the stator.

3.1 Yoke Thickness Variable with Variable Mass

The yoke thickness of the stator is variable and therefore the total mass of the stator is variable also.

In order to define the variable parameters, Fig. 3 shows a cross-sectional view of the stator. Parameter a, which is the tooth length, remained constant. Parameter b, which is the yoke thickness, is variable. The points on the outer radius are fixed. So, the outer radius is constant and the only variable is the middle radius.

The stator is simulated in 7 different cases where the middle radius varies in each simulation between 68 and 74 mm (sim1, sim2, etc. …). While the outer radius

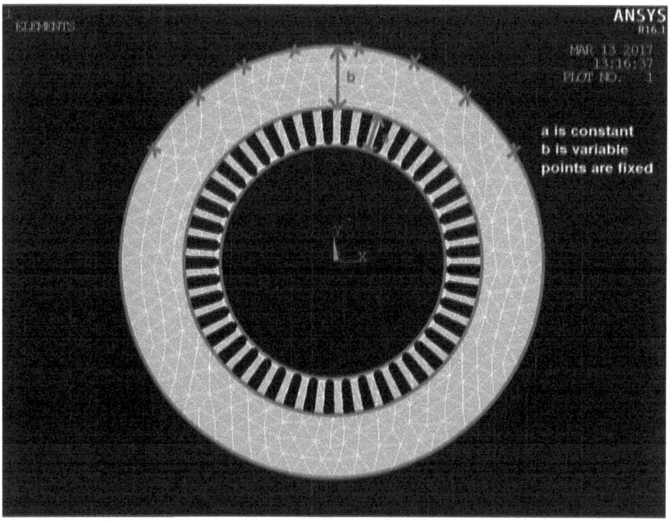

Fig. 3 Cross-sectional view of the stator

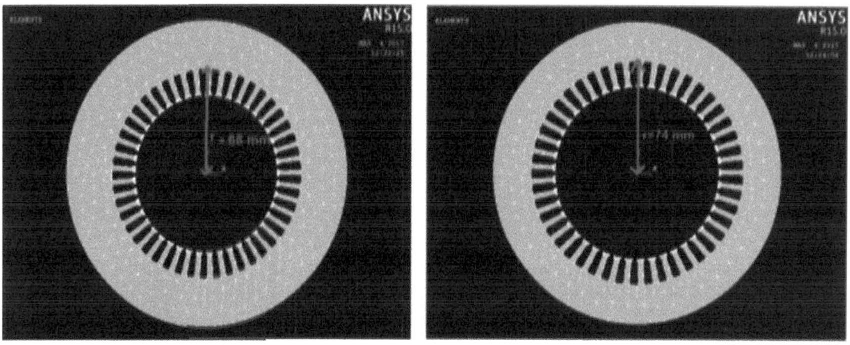

Fig. 4 Cross-sectional view of the stator in the first and last simulations

and the tooth length remain constant at 100 mm and 17 mm, respectively. Figure 4 presents the two extreme values of the stator middle radii. These 7 simulations were done using ANSYS-APDL using an implemented loop in the code text file.

The natural frequencies of the 7 simulations for mode 7, 21 and 39 are presented in Figs. 5, 6 and 7, respectively. These results are representative of all modes shape. In addition to the natural frequencies, these results show the mode shape of each mode.

As illustrated, the natural frequencies of the stator decrease as the middle radius increases. This can be related to the stiffness and the mass change in the stator. When the stator middle radius increases, the total mass of the stator decreases as well as the natural frequencies. This can lead us to conclude that the effect of the stator stiffness is more effective than the stator mass.

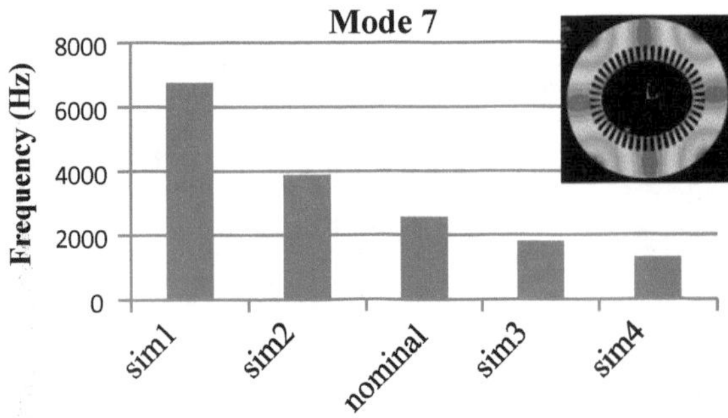

Fig. 5 Natural frequencies (Hz) of mode 7 in the first study

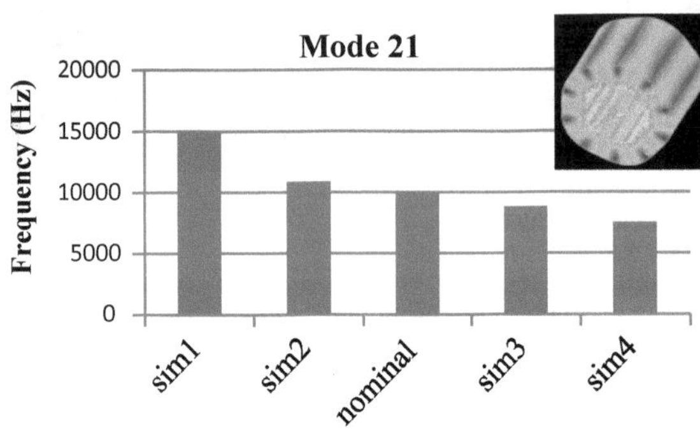

Fig. 6 Natural frequencies (Hz) for mode 21 in the first study

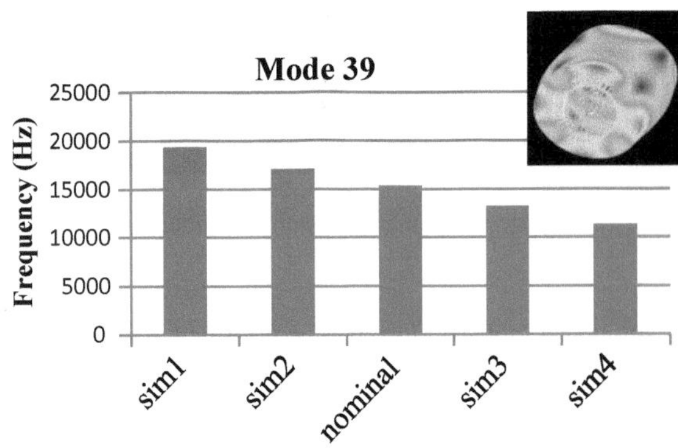

Fig. 7 Natural frequencies (Hz) for mode 39 in the first study

3.2 Yoke Thickness Variable with Constant Mass

In the second study, the yoke thickness of the stator is variable while the total mass of the stator remains constant. Figure 8 shows the cross-sectional view of the stator in order to define the variable parameters.

The main variable in this study is the yoke thickness (parameter b), the tooth length (parameter a) remains constant. In order to maintain the same mass while changing the yoke thickness, the outer radius will be variable here. However, the area of the stator will remain constant.

In this part, the FEM is applied in 5 cases where the yoke thickness and therefore the outer radius are variable. A pre-calculation is done in order to find a relationship between the outer radii and the middle one while keeping the area constant.

$$A_n = A_i \tag{1}$$

$$\pi\left(r_{on}^2 - r_{mn}^2\right) = \pi\left(r_{oi}^2 - r_{mi}^2\right) \tag{2}$$

Where:

- A: the cross section area of the stator
- r_o: the outer radius
- r_m: the middle radius
- The index n and i represent the nominal and the i simulation radius or area, respectively.

The outer radii r_{oi} varies between 80 and 120% of its nominal value r_{on}. Equation 3 represents the ratio between the simulation and the nominal outer radii.

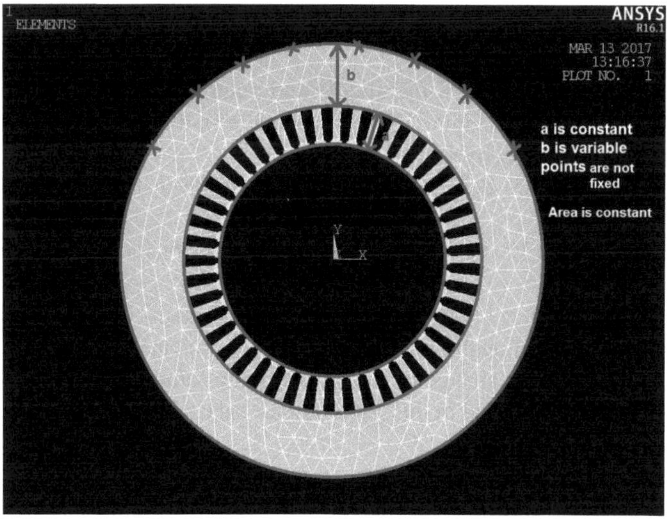

Fig. 8 Cross-sectional view of the stator in the second study

Table 3 Calculated values of the outer and middle radii (mm)

Simulation	Ratio (x)	r_{oi} (mm)	r_{mi} (mm)
1	0.8	80	37.96
2	0.9	90	56.04
3	1	100	71
4	1.1	110	84.5
5	1.2	120	97.16

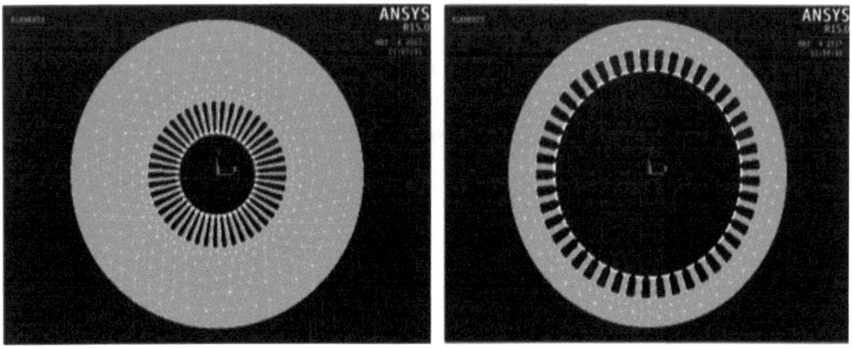

Fig. 9 Stator in the first (x = 0.8) and last (x = 1.2) simulations

$$r_{oi} = x r_{on} \tag{3}$$

Now, the middle radii can be expressed in Eq. 4 as a function of the simulated outer radii and the nominal outer and middle radii.

$$r_{mi} = \sqrt{\left(r_{oi}^2 - \left(r_{on}^2 - r_{mn}^2\right)\right)} \tag{4}$$

Table 3 shows the values of the middle and outer radii calculated using Eqs. 3 and 4 for each simulation. Figure 9 shows the stator in the first and last simulations (x = 0.8 and x = 1.2, respectively)

These parameters are defined and calculated in the text ANSYS-APDL finite element model code.

Figures 10, 11 and 12 show the natural frequencies obtained in the second study for the first modes.

It's clear that when x increases (outer and middle radii increased), the natural frequency decreases although the mass of the stator remains constant. Also, we can see that the effect of the change of x is globally the same for all the mode shapes (even for those not presented in this paper for the sake of brevity).

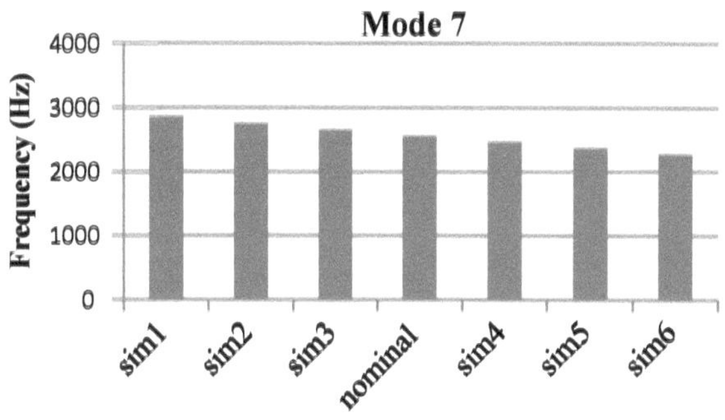

Fig. 10 Natural frequencies (Hz) for mode 7 in the second study

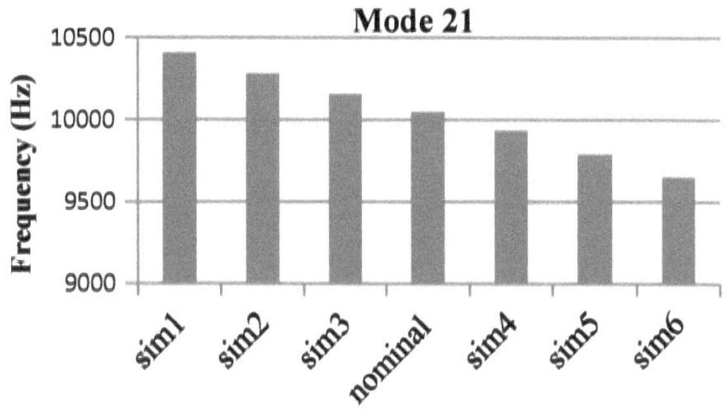

Fig. 11 Natural frequencies (Hz) for mode 21 in the second study

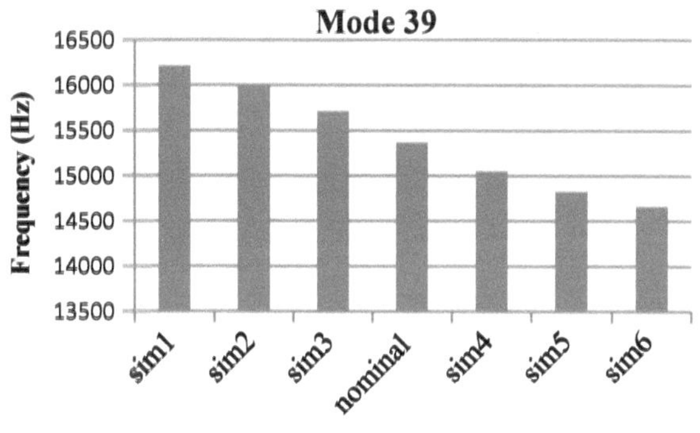

Fig. 12 Natural frequencies (Hz) for mode 39 in the second study

3.3 Analysis

The change in the yoke geometry affects the natural frequencies and the vibration behavior of the stator. Also, it is clear that the natural frequencies of the stator decrease as the yoke thickness is getting smaller although the mass remains constant or variable. Thus, one can conclude that the reduction of the yoke thickness increases the flexibility of the stator. On the other hand, the increase in the stator flexibility decreases the stiffness which is the main reason of the variation of the frequency.

The natural frequencies in the 2 studies presented decreases. This proves that the stator stiffness is more affected by changing the yoke geometry more than the mass.

4 Conclusion

In this paper, the effect of the stator geometrical change on the vibration behavior of an electric machine is presented. The stator of a wound rotor synchronous motor is chosen. The finite element software ANSYS APDL is used here due to its flexibility. The complete finite element model is created using APDL codes to manage and control the simulations. The convergence of the mesh was applied in order to get the optimal mesh. Using 6 nodes tetrahedral elements, the natural frequencies of the stator model are calculated in different cases, so that the effect of the geometry change on the vibration of the motor is evaluated with or without the effect of the mass.

As a sensitivity study, the natural frequencies of the stator are studied in 2 different cases for the first mode shapes. The 2 studies are related to the change of the yoke thickness of the stator. In the first study, the middle radius of the stator is changed in different values. While in the second study, the yoke thickness is changed while the total mass of the stator remained constant. In the 2 studies, the decrease in the yoke thickness leads to the decrease in the natural frequencies although the mass remained constant. As a conclusion, the decrease in the yoke thickness increases the stator flexibility which means that the stiffness is more effective than the mass.

References

1. Pellerey P, Lanfranchi V, Friedrich G (2012) Coupled numerical simulation between electromagnetic and structural models, influence of the supply harmonics for synchronous machine vibrations. IEEE Trans. Magn 48(2), Feb 2012
2. Hallal J, Pellerey P, Marion F, Druesne F, Lanfranchi V (2013) Harmonic pressure optimization on numerical electric motor model. In: COMPUMAG2013, Budapest, Hungary, July 2013
3. Druesne F, Hallal J, Lardeur P, Lanfranchi V (2016) Modal stability procedure applied to variability in vibration from electromagnetic origin for an electric motor. Finite Elements Anal Design 122:61–74
4. Boubaker MB, Druesne F, Lardeur P, Barillon F, Mordillat P (2012) Uncertain vibration analysis of an automotive car body modeled by finite elements with the modal statibility procedure. In: Proceedings of the international conference on uncertainty in structural dynamics (USD), Leuven, Belgium
5. More ST, Bindu RS (2015) Effect of mesh size on finite element analysis of plate structure. Int J Eng Sci Innov Technol (IJESIT) 4(3), May 2015

Part V
Mechanisms and Control

From Non-model-Based to Model-Based Control of PKMs: A Comparative Study

H. Saied, A. Chemori, M. El Rafei, C. Francis and F. Pierrot

Abstract This paper deals with control of parallel robots, where different controllers are proposed and compared. It demonstrates the strength of model-based controllers over the non-model-based ones when dealing with parallel kinematic manipulators known with their high nonlinearity, time-varying parameters and uncertainties. More precisely, adaptive model-based algorithms are the preferred control solutions for such kind of manipulators, thanks to their adjustable-parameters feature which is more adequate to the varying and non-accurate nature of parallel kinematic manipulators. These facts are fulfilled here by numerical simulations and real-time experiments on a four-degree-of-freedom parallel robot named VELOCE.

Keywords PKM · Dynamic model · Model-based · Non-model-based
Adaptive control · Feedforward · PID · Nonlinear PD · Augmented PD

H. Saied (✉) · A. Chemori · F. Pierrot
Laboratoire d'Informatique, de Robotique, et de Microélectronique de Montpellier, University of Montpellier, Montpellier, France
e-mail: hussein.saied@lirmm.fr

A. Chemori
e-mail: ahmed.chemori@lirmm.fr

F. Pierrot
e-mail: francois.pierrot@lirmm.fr

H. Saied · M. El Rafei · C. Francis
Scientific Research Center in Engineering, Lebanese University, Beirut, Lebanon
e-mail: maher.elrafei@ul.edu.lb

C. Francis
e-mail: cfrancis@ul.edu.lb

R. Rizk and M. Awad (eds.), *Mechanism, Machine, Robotics and Mechatronics Sciences*, Mechanisms and Machine Science 58,
https://doi.org/10.1007/978-3-319-89911-4_12

1 Introduction

Parallel kinematic manipulators (PKMs) are defined in [1] as follows: "A generalized parallel manipulator is a closed-loop kinematic chain mechanism whose end-effector is linked to the base by several independent kinematic chains".

PKMs were extensively used in robotized industries in the last few decades since it surpasses their counterpart's serial structures, particularly, in terms of high rigidity, better tracking performance, good precision, high payload-to-weight ratio and great dynamic [2, 3]. A very wide range of applications take benefit of PKMs. *Stewart* proposed *in 1965* a platform that is used as a flight simulator [4]. Delta robot prototype of 3 DOFs proposed initially in *1985* [5] is the leader in pick-and-place operations [1], used in packaging industry, laser cutting [6], medical applications [7] and haptic devices in which they allow the human-computer interaction [8]. Another parallel structures are used in machining tasks [9].

However, some problems associated with such kind of structures still open and are not solved satisfactory. The drawbacks of PKMs are listed as limited range of motion especially the rotational motion [10], small work space, low dexterity, complex forward kinematic solutions [11]. Singularities' behavior is more complicated than of serial [12], it can occur both inside and on the border of the work space [11].

In the literature, a wide range of control schemes have been proposed aiming to drive PKMs in accurate mode and high precision. The proposed control strategies can be classified in two classes, Model-Based and Non-Model-Based. The non-model-based strategies do not need a priori knowledge about the dynamics of the manipulator except the states (position and velocity). The Proportional-Integral-Derivative (PID) controller [13] is the most used in industrial applications mainly due to its simplicity and easy implementation as well as its acceptable control performance. However, PKMs known by their nonlinear dynamics, and highly increasing non-linearity at high speeds which may even lead to instability. The need for nonlinear controllers arises, knowing that PID lacks to robustness. Nonlinear PD (NPD) controller [14] is more adequate to the nature of PKMs, which can insure stability and disturbances rejection and performs with better robustness towards error variation. Successful application of non-model-based fuzzy controller applied on Stewart platform in [15], shows that this controller can drive the six-degree motion platform accurately, smoothly and in a stable way. On the other hand, researchers developed several model-based controllers depending on the fact that the closed-loop algorithms, rich enough with knowledge about the system dynamics, can compensate their nonlinearities. PD with gravity compensation or with desired gravity compensation were applied intending to achieve better performance than simple PD since it surpasses the effect of gravity [16, 17]. Computed torque (CT) control exploits the full knowledge about the nonlinear system dynamics, leading to a linear closed-loop system in terms of tracking error [18]. Also the Augmented PD (APD) is a model-based strategy, where the dynamic part of the controller is computed from both the desired and measured states improving the global performance of the control mission [19]. Nevertheless, PKMs are featured with time-varying parameters (e.g. payload mass), uncertainties and difficulty to get accurate model values, then the design of adaptive controllers is very significant. Adaptive model-based controllers recompense the possible variation of parameters and react against the disturbances by dynamical calibration in an online algorithm, such as the adaptive feedforward PD controller (AFFPD) [20].

The control performances of some classical non-model-based controllers, as PD, PID, NPD, and model-based controllers, as APD, AFFPD, are studied and compared in this paper. The main objective is to show that a controller fed with a good dynamic knowledge about the robot will be very powerful and more precise. Moreover, the time-varying PKM environment requires adaptive dynamic knowledge to manage robustness and accuracy, as it was proved with real-time experimental tests.

The paper organization is as follow: Sect. 2 describes the structure of VELOCE parallel robot, as well as its kinematic and dynamic modeling. Section 3 is dedicated to the synthesis of the proposed control solutions. Simulation and experimental results are presented and discussed in Sect. 4. Section 5 concludes the paper and states the future work.

2 Description and Modeling of VELOCE PKM

In this section, a full description of the whole mechanical structure of VELOCE PKM is presented, then a brief explanation of its kinematic and dynamic models is introduced.

2.1 Structure of VELOCE PKM

VELOCE robot (see Fig. 1) is a 4-DOF parallel manipulator designed and fabricated in LIRMM (Laboratoire d'Informatique, de Robotique et de Microélectronique de Montpellier). It is mainly designed for pick-and-place applications. It consists of four kinematic chains and four degrees-of-freedom, three independent translational degrees in the three dimensions and one rotational degree around the vertical axis. It is note that VELOCE is a fully parallel manipulator [1]. Each kinematic chain is composed, in a serial manner, of an actuator, a rear arm fixed to the actuator's rotor, a forearm including two links forming a parallelogram and connected through ball joints to the rear arm and to the traveling plate (see Fig. 2). The traveling plate is made of two essential parts, upper and lower. Both parts are mounted on a single screw, and the movement of one part with respect to the other generates the rotational action.

Fig. 1 VELOCE PKM. **a**: CAD view, **b**: The manufactured robot

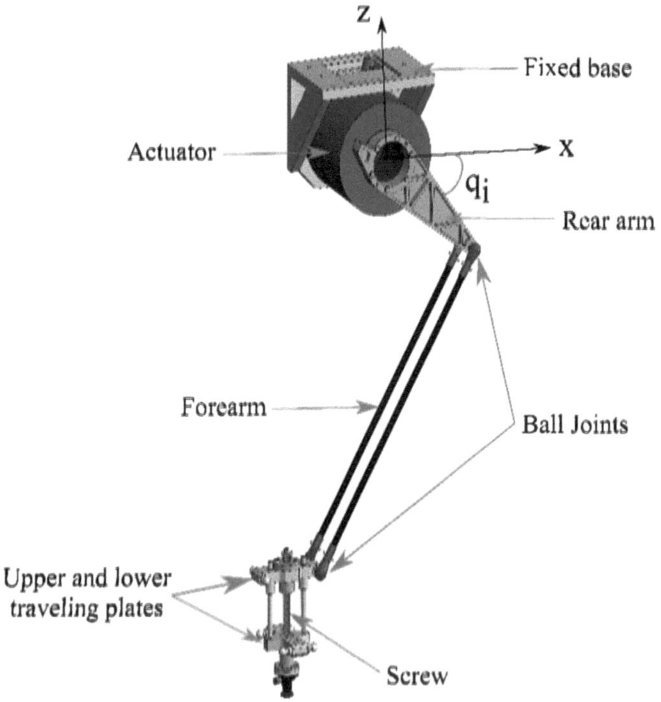

Fig. 2 An *i*th kinematic chain of VELOCE PKM

2.2 Kinematic Modeling of VELOCE PKM

The Cartesian coordinates of the traveling plate can be presented with respect to the fixed-base frame in four-dimensional space vector $\boldsymbol{x} = [x, y, z, \alpha]^T$ such that x, y, z are the translational coordinates and α is the rotational angle around z-axis. The orientation and position of the traveling plate are specified by the angular positions of the four actuators, since VELOCE is a fully PKM, represented in another four-dimensional space vector $\boldsymbol{q} = [q_1, q_2, q_3, q_4]^T$. The relation between $\boldsymbol{q}$ and $\boldsymbol{x}$ is obtained by a geometrical study for the constraints of the closed-loop formed of kinematic chains and traveling plate. The study leads to the following kinematic models, "Forward kinematic (Fk)" and "Inverse kinematic (Ik)" respectively: $\boldsymbol{x} = Fk(\boldsymbol{q})\,; \boldsymbol{q} = Ik(\boldsymbol{x})$. Applying the *equiprojectivity principle* explained in [3], the inverse Jacobian matrix can be computed and thus a relation between the joints' velocities and Cartesian velocity of traveling plate is formulated as follows:

$$\dot{q} = \boldsymbol{J}_m \dot{x} \tag{1}$$

where $\boldsymbol{J}_m \in \mathbb{R}^{4\times 4}$ is the inverse Jacobian Matrix. Note that $\boldsymbol{J}_m$ is square and invertible for the fully PKMs (as VELOCE), and if the chosen trajectory is away from singularities. By differentiating Eq. (1) with respect to time, we obtain the relation of accelerations between Cartesian space and joint space as follows:

$$\ddot{q} = \boldsymbol{J}_m \ddot{x} + \dot{\boldsymbol{J}}_m \dot{x} \tag{2}$$

2.3 Dynamic Modeling of VELOCE PKM

According to [21], the dynamic model can be obtained by analyzing the dynamics in the joint space and in the traveling-plate space separately, then summing up the two equations of motion. Nonetheless, some assumptions are taken to simplify the complexity of the rigid body of such robots. Standing on the light weight of the forearm, its rotational inertia is neglected and its mass is split-up into two parts, one part conjoined to the rear arm and one part to the traveling plate mass. Also the dry and viscous friction in the passive and active joints are ignored, and the effect of gravity can be omitted at high speeds. Regarding the traveling plate, there are three kind of forces acting on it: the gravity forces, the inertial forces and the forces of the load. These forces are transformed into contributions in actuators' torques using the Jacobian matrices. From the joints side, the gravity of the rear arms with the half-masses of forearms and the arms inertia are also expressed in the actuators' torques. Then, the total actuators' torques vector is computed by summing up the contributions of all forces. One can formulate the total inverse dynamic equation to be in the standard joint space form, so we get:

$$\boldsymbol{M}(\boldsymbol{q})\ddot{\boldsymbol{q}} + \boldsymbol{C}(\boldsymbol{q},\dot{\boldsymbol{q}})\dot{\boldsymbol{q}} + \boldsymbol{G}(\boldsymbol{q}) + \boldsymbol{\Gamma}_{F_{load}} = \boldsymbol{\Gamma} \quad (3)$$

with $\boldsymbol{M}(\boldsymbol{q}) \in \mathbb{R}^{4\times4}$ being the inertia matrix, $\boldsymbol{C}(\boldsymbol{q},\dot{\boldsymbol{q}}) \in \mathbb{R}^{4\times4}$ is the Coriolis and centrifugal forces matrix, $\boldsymbol{G}(\boldsymbol{q}) \in \mathbb{R}^4$ be the gravitational forces vector and $\boldsymbol{\Gamma}_{F_{load}} \in \mathbb{R}^4$ be the payload forces vector. A fundamental property of PKMs is very essential for model-based adaptive controllers consists of linearity of the dynamics with respect to the parameters, such as inertia and masses [22]. So the reformulation of the dynamics in the linear form is expressed as following:

$$\boldsymbol{Y}(\boldsymbol{q},\dot{\boldsymbol{q}},\ddot{\boldsymbol{q}})\boldsymbol{\Phi} = \boldsymbol{\Gamma} \quad (4)$$

where $\boldsymbol{Y}(.) \in \mathbb{R}^{4\times n}$ is the regression matrix which is nonlinear function in terms of $\boldsymbol{q}, \dot{\boldsymbol{q}}$ and $\ddot{\boldsymbol{q}}$, and $\boldsymbol{\Phi} \in \mathbb{R}^n$ is the robot parameters vector to be estimated.

3 Proposed Control Solutions

In this section, a design for the proposed control solutions is clarified. The controllers designed are non-model-based controllers: PD, PID and NPD, and model-based controllers: APD and AFFC. Note that the available measurements are directly the joint angles of the actuators, so all controllers are developed in joint space.

3.1 *Proportional-Derivative Controller*

PD control scheme is composed of two parts, proportional and derivative parts. The general expression of the control input is:

$$\boldsymbol{\Gamma} = \boldsymbol{K}_p\boldsymbol{e} + \boldsymbol{K}_d\dot{\boldsymbol{e}} \quad (5)$$

where $\boldsymbol{e} = \boldsymbol{q}_d - \boldsymbol{q}$ is the joint position error between the desired angles and the actual measured ones. $\boldsymbol{K}_p, \boldsymbol{K}_d \in \mathbb{R}^{4\times4}$ are diagonal positive definite matrices which means that no coupling between the joints is considered, and the controller is called a linear single-axis controller [2], knowing that the same gain is used for all joints. The PD control law is asymptotically stable as was addressed in [22]. It is the simplest control law but it has several drawbacks briefed as weak disturbance rejection, no compensation for the nonlinearity and variation nature, and even may leads to instability at high accelerations.

3.2 Proportional-Integral-Derivative Controller

It is the same demonstration of the aforementioned PD controller just adding the integral term which is the multiplication of the integral of position error with a positive constant feedback. The control law equation is then:

$$\boldsymbol{\Gamma} = \boldsymbol{K}_p\boldsymbol{e} + \boldsymbol{K}_i \int \boldsymbol{e}\,dt + \boldsymbol{K}_d\dot{\boldsymbol{e}} \tag{6}$$

where $\boldsymbol{K}_i \in \mathbb{R}^{4\times4}$ is a diagonal positive definite matrix. Same specifications and draw backs of the PD control, but better global performance related to the tracking error thanks to the contribution of integral term in eliminating the residual errors in the steady state response produced by the proportional term.

3.3 Nonlinear Proportional-Derivative Controller

This controller have the same structure of classical PD controller with the time-varying feedback gains instead of being constant. The feedback gains are nonlinear functions in terms of the system states, inputs, and other variables. As For the classical PD, the control law equation can be written as following with the nonlincar gains functions [3]:

$$\boldsymbol{\Gamma} = \boldsymbol{K}_p\boldsymbol{f}\left(\boldsymbol{e}, \alpha_1, \delta_1\right)\boldsymbol{e} + \boldsymbol{K}_d\boldsymbol{f}\left(\dot{\boldsymbol{e}}, \alpha_2, \delta_2\right)\dot{\boldsymbol{e}} \tag{7}$$

Where

$$\boldsymbol{f}\left(\boldsymbol{x}, \alpha, \delta\right) = \begin{cases} |\boldsymbol{x}|^{\alpha-1}, & |\boldsymbol{x}| > \delta \\ \delta^{\alpha-1}, & |\boldsymbol{x}| \le \delta \end{cases} \tag{8}$$

with α_1 and α_2 can be chosen within the interval [0.5, 1] and [1, 1.5] respectively. δ_1 and δ_2 are positive constant numbers. From the above nonlinear structure, the feedback gains are adjusted online depending on the value of the error. For small position error, a large gain is produced, and for large position error, a small gain is obtained. On the other hand, large gains for large error rate and small gains for small error rate. This behavior results with rapid transition of the system and favorable damping. NPD is a robust controller against the nonlinearities of PKMs, parametric uncertainties and time delays.

3.4 Augmented Proportional Derivative Controller

APD, known also as PD+, is one of the conventional model-based controllers composed of two main parts, feedback part and dynamic model part. The feedback part is a simple PD controller that guarantees the stability and the dynamical part represents the nonlinear dynamics of the system that compensates its effects and enhances the control performance. The control law form of APD looks as follows [19]:

$$\boldsymbol{\Gamma} = \boldsymbol{M}(\boldsymbol{q})\,\ddot{\boldsymbol{q}}_d + \boldsymbol{C}(\boldsymbol{q},\dot{\boldsymbol{q}})\,\dot{\boldsymbol{q}}_d + \boldsymbol{G}(\boldsymbol{q}) \\ + \boldsymbol{\Gamma}_{F_{load}} + \boldsymbol{K}_p\boldsymbol{e} + \boldsymbol{K}_d\dot{\boldsymbol{e}} \tag{9}$$

As shown in Eq. (9), the dynamical term is computed from the desired and actual trajectories. However, such kind of controllers relying mostly on the dynamics of the robot needs to have an accurate model information, and it cannot compensate the effect of time-varying parameters and uncertainties of PKMs.

3.5 Adaptive Feedforward with PD Controller

The AFFPD controller is quietly similar to the APD in the general form, meaning that it is divided into two parts, one part a simple PD feedback to conserve the stability and the other part is the adaptive feedforward dynamics of the PKM to reduce the influence of variation in parameters and uncertainties. Thanks to the property of linearizing the dynamic model, the adaptive term is the multiplication of the regression matrix with the estimated vector of parameters [20]. The control law equation is as follows:

$$\boldsymbol{\Gamma} = \boldsymbol{Y}\left(\boldsymbol{q}_d, \dot{\boldsymbol{q}}_d, \ddot{\boldsymbol{q}}_d\right)\widehat{\boldsymbol{\Phi}} + \boldsymbol{K}_p\boldsymbol{e} + \boldsymbol{K}_d\dot{\boldsymbol{e}} \tag{10}$$

All the parameters $\left(\widehat{\boldsymbol{\Phi}} \in \mathbb{R}^6\right)$ need to be estimated and adapted depending on the error. The controller relies on the desired trajectories instead of the measured ones which can improve the efficiency. The estimating algorithm is in function of the measured error as follows:

$$\dot{\widehat{\boldsymbol{\Phi}}} = \boldsymbol{K}\boldsymbol{Y}(.)^T\,\boldsymbol{\tau}_{FB} \tag{11}$$

where $\boldsymbol{K} \in \mathbb{R}^{6\times 6}$ is a positive definite matrix that need to be chosen for a good estimation and tracking error stability. $\boldsymbol{\tau}_{FB}$ is the torque computed from the feedback part. After linearizing the inverse dynamic Eq. (3) considering all the PKM parameters need to be estimated, the following vector of parameters is obtained: $\Phi = [MTP\ MUTP\ Ms\ I_a\ m_a r_a\ m_{load}]^T$ such that MTP is the total mass of the traveling plate including the contribution of the forearms, $MUTP$ is the mass of the upper traveling plate including the contribution of the forearms, Ms is mass of upper traveling plate with the equivalent mass to rotate the screw. According to [23], the used adaptive control scheme achieve a global asymptotic stability respecting the necessary and sufficient conditions for adaptive control [24], in which the reference trajectory should be chosen rich enough with frequencies to converge the parameters estimation's error to zero, with a suitable initial values of the parameters.

4 Numerical Simulations and Experiments

VELOCE has four direct-drive motors TMB0140-100-3RBS ETEL, they can provide maximum torque of 127 Nm and reach up a speed of 550 rpm. All actuators are supplied with non-contact incremental optical encoders of 5000 pulses per revolution. The global structure can hold as maximum payload of 10 kg, achieve a peak velocity of 10 m/s and peak acceleration of 200 m/s^2.

4.1 Simulation Results

Simulations were done in Matlab/Simulink environment implementing the controllers in discrete-time schemes similar to real robots control. A fixed-step solver was chosen of sample-time equal to 0.1 ms. The chosen desired trajectory is a sequence of point-to-point motions with a duration of each motion T=0.5 s. A nominal scenario of motion is used to compare the performances of the controllers such that no payload is considered in the simulations. The evaluation criteria proposed to monitor the performances is the computation of the Root Mean Square Error over the Translational (RMSET) and Rotational (RMSER) degrees-of-freedom as follow:

$$RMSET = \left(\frac{1}{N}\sum_{i=1}^{N}\left(e_x^2(i) + e_y^2(i) + e_z^2(i)\right)\right)^{\frac{1}{2}} \tag{12}$$

$$RMSER = \left(\frac{1}{N}\sum_{i=1}^{N}\left(e_\alpha^2(i)\right)\right)^{\frac{1}{2}} \tag{13}$$

where N is the number of the time-samples, e_x, e_y, e_z represent the tracking errors along the axes x, y and z, e_α denotes the tracking error along denotes the tracking error along the rotational angle. The gains for each controller in these simulations are specified by the trial and error technique and shown in Table 1. The comparison between the three non-model-based controllers, in Fig. 3, shows that a NPD performs better than the linear controllers (PD, PID). Thanks to its adjustable gains with the error state, as discussed before, that grants it more robustness and rejection for nonlinearity. For clarity, a zoom in from 4 to 6 s is done in the plot of the Cartesian error in Fig. 3, and the control input signals for the three controllers are depicted

Table 1 Control design gains in numerical simulation

PD/APD	PID	NPD	AFFPD
$k_p = 4000$ $k_d = 6$	$k_p = 4000$ $k_d = 6$ $k_i = 500$	$k_p = 2800$ $\alpha_1 = 0.5$ $\delta_1 = 0.0062$ $k_d = 10$ $\alpha_2 = 1.5$ $\delta_2 = 2.4131$	$k_p = 8000$ $k_d = 100$ $K = diag\left(\left[100\,100\,5 \times 10^4\,0.5\,0.51\right]\right)$

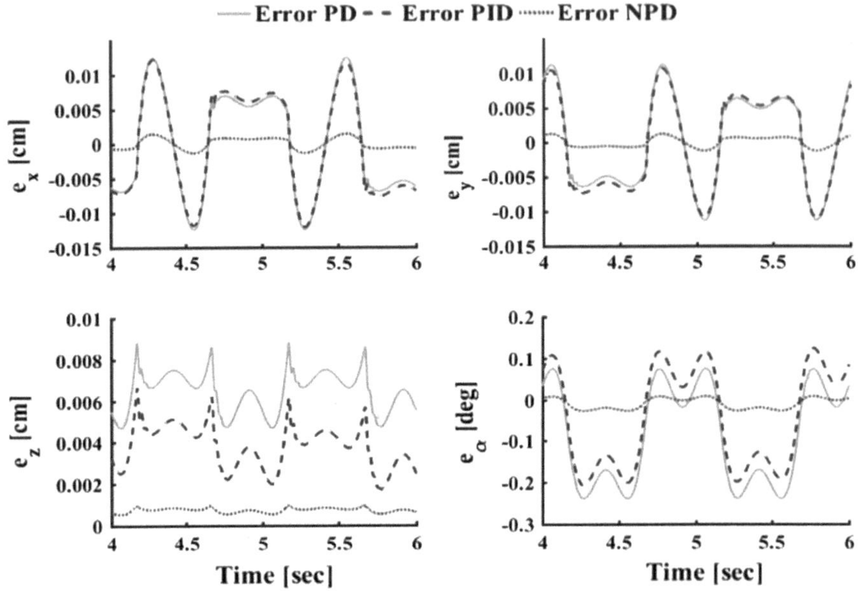

Fig. 3 Evolution of the Cartesian tracking error in numerical simulation

in Fig. 4. It is obvious that the control input signals are within the allowable range that can be handled by the real actuators. Similarly, the comparison of the moving platform's tracking error for the two model-based controllers is presented in Fig. 5. Apparently, the benefit of parameters' adaptation in the closed-loop of a controller (AFFPD) is very significant in improving the precision and accuracy, unlike the non-adaptive model-based controller (APD) which is limited in rejecting the uncertainties and parameters variation. Both control signals are still under saturation and proper with the real actuators limits (see Fig. 6). A good parameters' estimation convergence of the AFFPD controller is shown in Fig. 7 reducing more the moving platform's tracking error, knowing that we initialize the parameters with much closed values to the optimal numbers. The quantifications of the errors all over the trajectory are shown in Table 2 with the improvements of each controller. It is notable to say that the simulated model-based controllers are more accurate than the non-model-based, as the calculation of the percentages shows high improvements.

Table 2 Performance evaluation of the controllers in numerical simulation

	PD	PID	Improvement on PD (%)	NPD	Improvement on PID (%)	APD	Improvement on NPD (%)	AFFPD	Improvement on APD (%)
RMSET [cm]	0.0094	0.0078	17.02	0.0011	85.9	2.1018×10^{-5}	98.09	1.0862×10^{-5}	48.32
RMSER [deg]	0.1309	0.0983	24.90	0.0147	85.04	1.6907×10^{-4}	98.85	5.309×10^{-5}	68.6

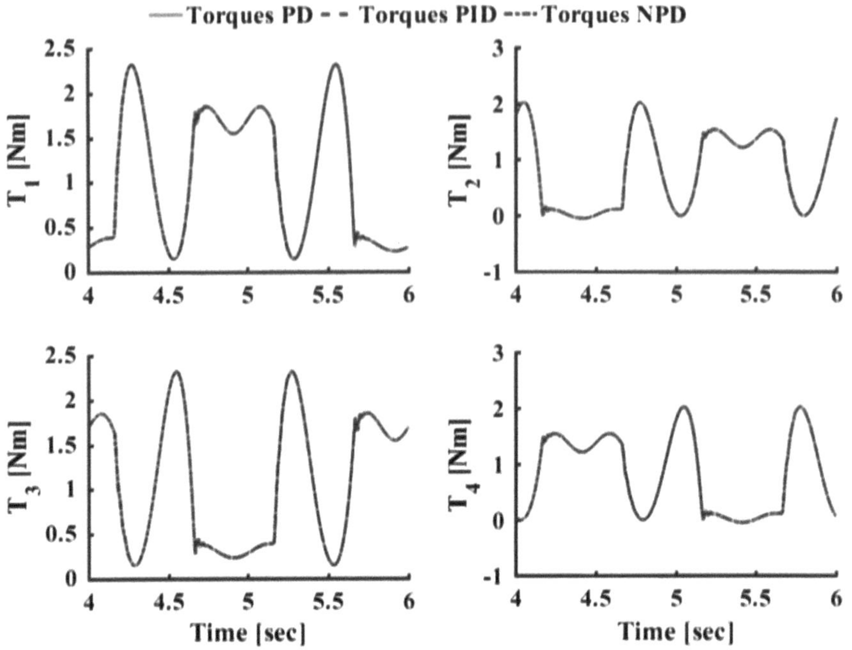

Fig. 4 Evolution of the control input signals in numerical simulation

4.2 Experimental Results

Due to its interesting specifications, AFFPD controller is applied in real-time experiments on VELOCE robot, and compared to the PD controller. The control architecture of the VELOCE robot is implemented using Simulink from Mathworks Inc. and compiled using XPC Target (an industrial computer of frequency 10 kHz i.e. the sample time is 0.1 ms) and the Real-Time toolboxes. Same evaluation criteria used in simulations is considered in the experiments. Retuning the gains of control design is needed for experiments, and the obtained gains for AFFPD are: $k_p = 4000, k_d = 6$, and $K = diag\left(\left[2.5 * 10^{-3}\ 0.125\ 0.1\ 10^{-5}\ 10^{-5}\ 10^{-3}\right]\right)$. The plot of the tracking error in Cartesian space for both controllers is represented in Fig. 8, showing the better and improved global performance of AFFPD controller with respect to PD controller. The evaluations and improvements in the tracking error are computed and shown in Table 3 validating our pretend that model-based controllers are more powerful than non-model-based controllers in real-time experiments. More precisely, control schemes that include adaptive dynamics provide robustness against parameters variation and uncertainties. The control input signals of both controllers are under saturation and in the safe range (see Fig. 9). Figure 10 visualize a good convergence for the estimated parameters in the AFFPD controller, which contributes in minimizing the tracking error as possible. One can notice the

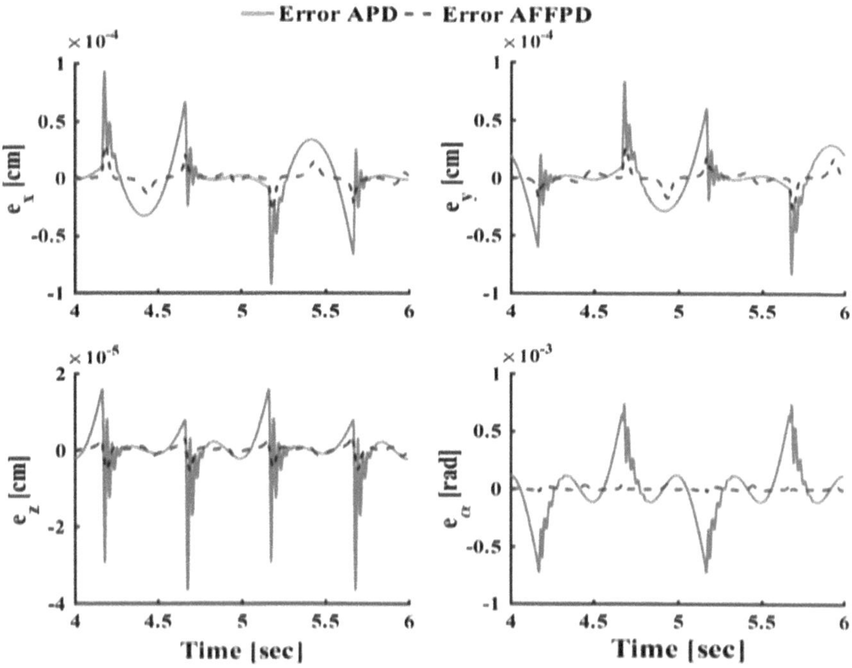

Fig. 5 Evolution of the Cartesian tracking error in numerical simulation

Table 3 Performance evaluation of PD and AFFPD controllers in real-time experiments

	PD	AFFPD	Improvements (%)
RMSET [cm]	0.0156	0.0092	41.03
RMSER [deg]	1.077	0.7596	29.47

degradation of such root mean squares of Cartesian tracking error of the two controllers from numerical simulations to real-time experiments, and that is normal because of the inaccurate model of PKMs exist in the literature, in which they simplify friction, actuators' dynamics, transmission system,…etc.

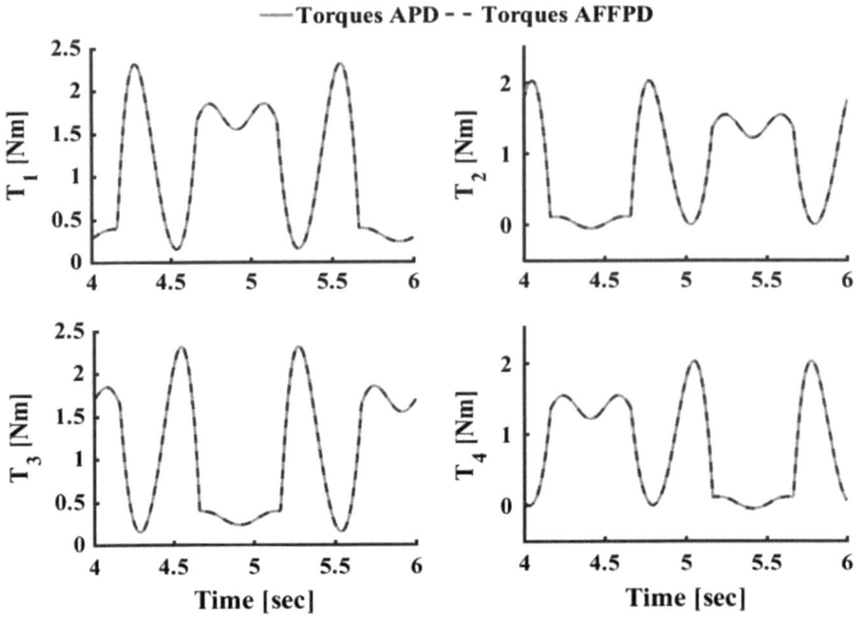

Fig. 6 Evolution of the control input signals in numerical simulation

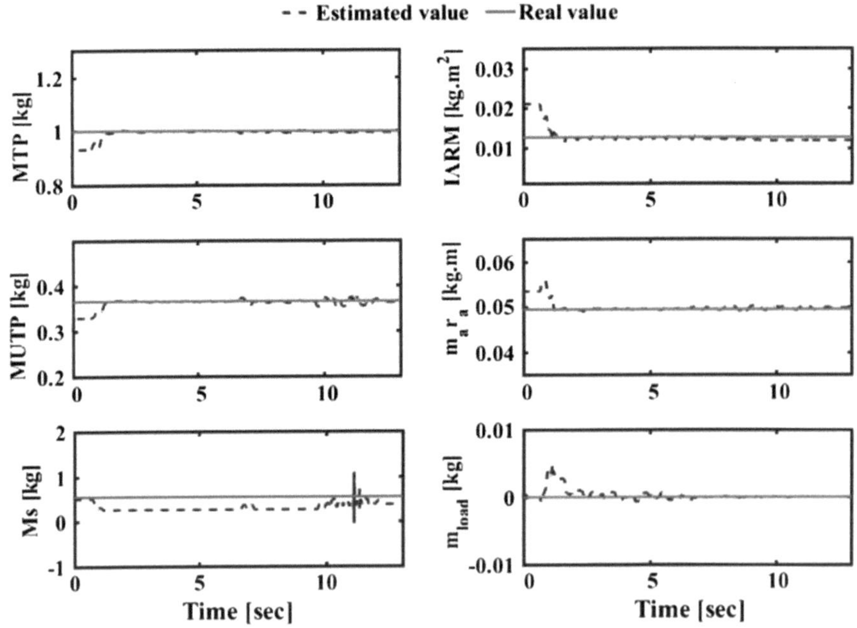

Fig. 7 Parameters estimation in numerical simulation of AFFPD

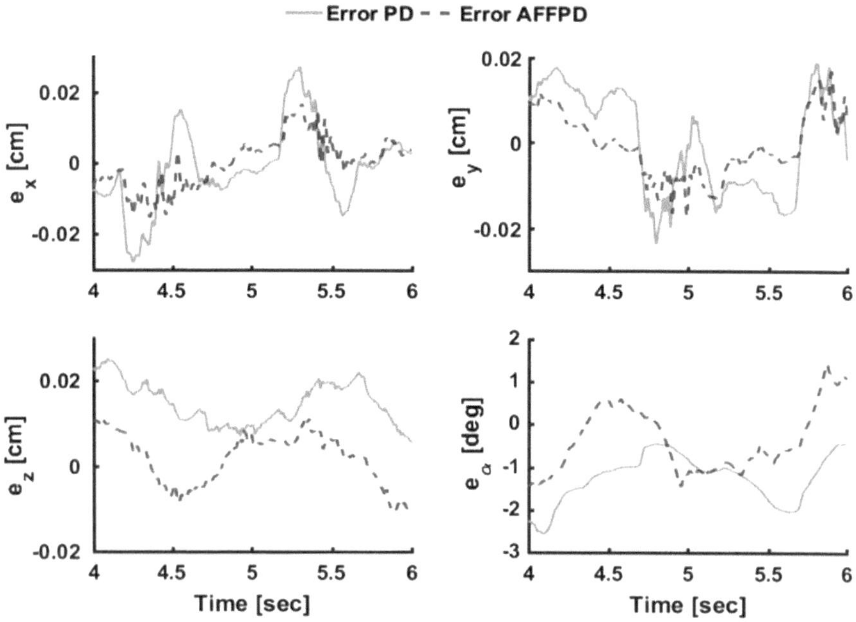

Fig. 8 Evolution of the Cartesian tracking error in real-time experiments

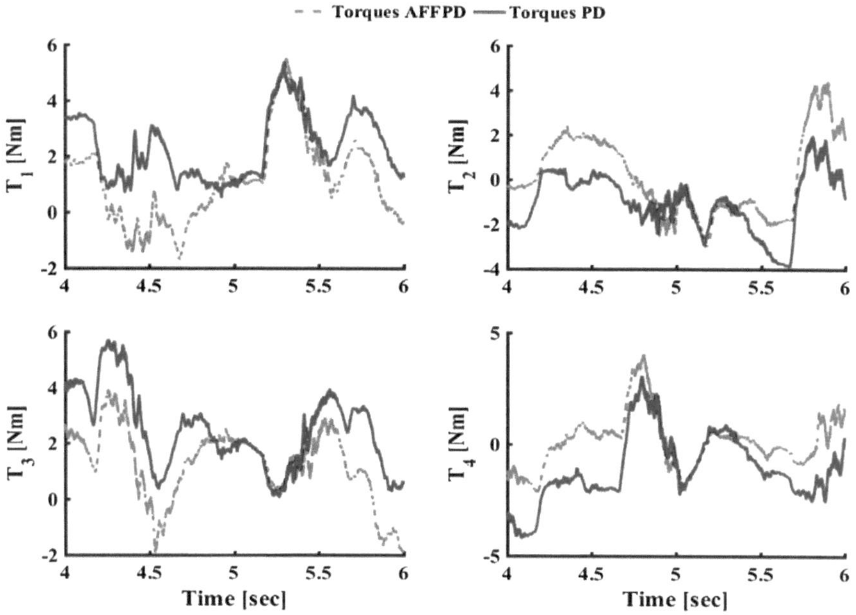

Fig. 9 Evolution of the control input signals in real-time experiments

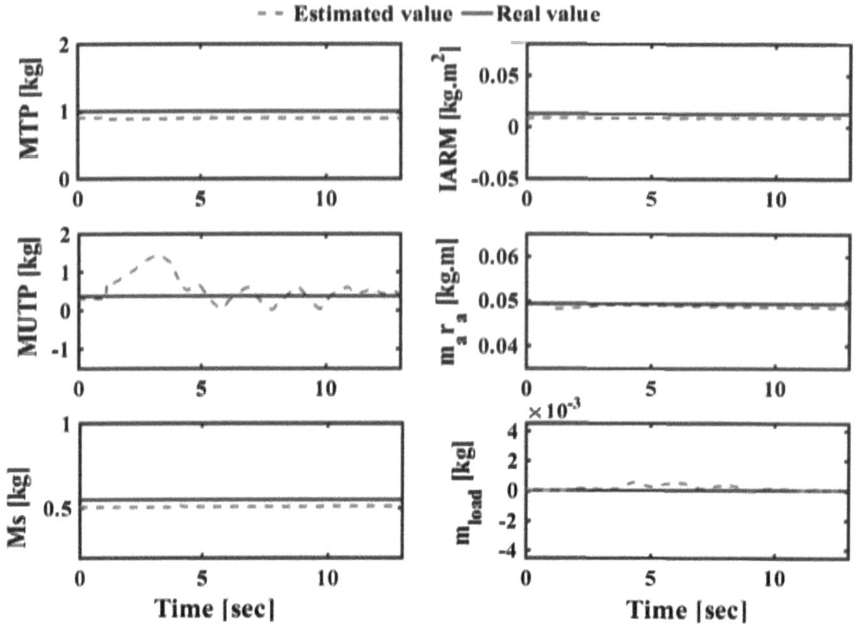

Fig. 10 Parameters estimation in real-time experiments of AFFPD

5 Conclusion and Future Work

In this paper, a comparison between the performances of model-based (Augmented PD, Adaptive Feedforward with PD) and non-model-based (PD, PID, Nonlinear PD) controllers was done by numerical simulation sketching and interpreting. We show the importance of including the dynamic model of the PKMs in the closed-loop control, and its main role in enhancing the performance of the controller, especially when adapting the dynamical parameters of the PKMs. Real-time experiments of PD and AFFPD controllers were conducted on a 4-DOF parallel robot to verify the validation of simulation results in the real applications of parallel robots.

As a future perspective, one can look for more accurate models of PKM involve the full dynamics such as articulations' friction, actuators' dynamics, motor drivers, and transmission system. Corporate these models in adaptive closed-loop algorithms to improve the performance of parallel robots, in terms of precision, motion speed and robustness.

Acknowledgements This paper has been supported by the Erasmus+ mobility project, ARPE ARROW project and the Lebanese University.

References

1. Merlet J-P (2006) Parallel robots. Springer, The NetherLands
2. Bennehar M (2015) Some contributions to nonlinear adaptive control of PKMs : from design to real-time experiments. PhD thesis, University of Montpellier
3. Natal GS (2016) Control of parallel robots: towards very high accelerations. PhD thesis, University of Montpellier
4. Stewart D (1965) A platform with six degrees of freedom. Arch Proc Inst Mech Eng 1847–1982(180):371–386
5. Clavel R (1990) Device for the movement and positioning of an element in space, Switzerland. US Patent 4976582A. https://patents.google.com/patent/US4976582A
6. Bruzzone L, Molfino R, Razzoli R (2002) Modelling and design of a parallel robot for laser-cutting applications. In: Proc. IASTED international conference on modeling, identification and control, Innsbruck, Austria
7. Yangmin L, Qingsong X (2007) Design and development of a medical parallel robot for cardiopulmonary resuscitation. IEEE/ASME Trans Mechatron 12:265–273
8. Grange S, Conti F, Rouiller P, Helmer P, Baur C (2001) Overview of the delta haptic device. Proc Eurohaptics 1:5–7
9. Toyama T, Susuki M (1998) Machine tool having parallel structure, Japan. US Patent 5,715,729, 1998. https://www.google.com/patents/US5715729
10. Kucuk S (2012) Serial and parallel robot manipulators-kinematics, dynamics, control and optimization. InTech, Croatia
11. Simaan N (1999) Analysis and synthesis of parallel robots for medical applications. Master thesis, Israel Institute of Technology
12. Liu GF, Wu YL, Wu XZ, Kuen YY, Li ZX (2001) Analysis and control of redundant parallel manipulators, South Korea. In: Proceedings of 2001 ICRA, IEEE international conference on robotics and automation (Cat. No. 01CH37164), vol 4, pp 3748–3754
13. Ziegler JG, Nichols NB (1942) Optimum settings for automatic controllers. Trans Am Soc Mech Eng 64:759–768
14. Su YX, Duan BY, Zheng CH (2004) Nonlinear PID control of a six-DOF parallel manipulator. IEE Proc Control Theory Appl 151:95–102
15. Chung I-F, Chang H-H, Lin C-T (1999) Fuzzy control of a six-degree motion platform with stability analysis. In: IEEE SMC'99 Proceedings of 1999 IEEE international conference on systems, man, cybernetics (Cat. No. 99CH37028), vol 1, pp 325–330
16. Yang C, Huang Q, Jiang H, Ogbobe Peter O, Han J (2010) PD control with gravity compensation for hydraulic 6-DOF parallel manipulator. Mech Mach Theory 45:666–677
17. Kelly R (1997) PD control with desired gravity compensation of robotic manipulators: a review. Int J Rob Res 16:660–672
18. Codourey A (1998) Dynamic modeling of parallel robots for computed-torque control implementation. Int J Rob Res 17:1325–1336
19. Shang WW, Cong S, Li ZX, Jiang SL (2009) Augmented nonlinear PD controller for a redundantly actuated parallel manipulator. Adv Robot 23:1725–1742
20. Codourey A, Honegger M, Burdet E (1997) A body-oriented method for dynamic modeling and adaptive control of fully parallel robots. In: Proceedings of 5th Symposium Robot Control, pp 443–450
21. Nabat V (2007) Robots parallèles à nacelle articulée, du concept à la solution industrielle pour le pick-and-place. PhD thesis, University of Montpellier
22. Khalil W, Dombre E (2004) Modeling, identification and control of robots. Elsevier Ltd., London
23. Sadegh N, Horowitz R (1990) Stability and robustness analysis of a class of adaptive controllers for robotic manipulators. Int J Rob Res 9:74–92
24. Boyd S, Sastry SS (1986) Necessary and sufficient conditions for parameter convergence in adaptive control. Automatica 22:629–639

Parallel Robot Integration and Synchronization in a Waste Sorting System

E. Mokled, G. Chartouni, C. Kassis and R. Rizk

Abstract This paper presents a delta robot used in a waste sorting system, where its role is to separate plastics and glass from the main waste line. A special detector gives the position and the timing. The trajectory of the picked parts respects the continuity of velocity and acceleration. The pick and place algorithm aims to reach the maximum number of removed items per minute. The inverse kinematics model helps to find the trajectory in the joint space as well as the maximum value of the acceleration. Finally a prototype of the model is presented.

Keywords Parallel mechanisms · Kinematic models · Pick and place
Waste sorting

Nomenclature

$\{B\}$	fixed base Cartesian reference frame
$\{P\}$	moving platform Cartesian reference frame
${}_{i}^{R}R_Z$	Transformation matrix between frames $\{R_i\}$ and $\{R\}$
θ_i	Angle of the first upper arm attached to motor i
$\dot{\theta}_l$	angular velocity of the upper arm i
x, y, z	Cartesian variables of the platform center
$\dot{x}, \dot{y}, \dot{z}$	velocity variables of the platform center

E. Mokled · G. Chartouni · C. Kassis (✉)
Faculty of Engineering, CRSI, Lebanese University, Beirut, Lebanon
e-mail: carine.kassis@ul.edu.lb

E. Mokled
e-mail: eliasmoukled@gmail.com

G. Chartouni
e-mail: georges2222@hotmail.com

R. Rizk
Faculty of Engineering, CRSI, Lebanese University, Roumieh, Lebanon
e-mail: rrizk2@ul.edu.lb

R. Rizk and M. Awad (eds.), *Mechanism, Machine, Robotics and Mechatronics Sciences*, Mechanisms and Machine Science 58,
https://doi.org/10.1007/978-3-319-89911-4_13

l	length of arm
L	length of forearm
v_c	velocity of the conveyor
B_i	hips of leg i
A_i	knee of leg i
P_i	ankles of leg i
w_B	planar distance from $\{B\}$ to near base side
u_B	planar distance from $\{B\}$ to a base vertex
w_p	planar distance from $\{P\}$ to near platform side
u_p	planar distance from $\{P\}$ to a platform vertex
s_B	size of the base side
s_p	size of the moving platform side

1 Introduction

Parallel mechanisms are emerging in industry (for instance machine-tools, high-speed pick-and-place robots, flight simulators, medical robots). A parallel mechanism can be defined as a mechanism with a closed kinematic chain. It is made up of an end-effector with N degrees of freedom and a fixed base connected to each other by at least two kinematic chains, the motorization being carried out by N actuators [1]. This allows parallel mechanisms to bear higher loads at higher speed and often with a higher repeatability [1]. Waste sorting is a pick and place task. In this case, high speed manipulation is needed but the payload is not heavy. Thus, parallel manipulator represents an ideal solution. However, with the high speed and accelerations, inertia becomes an important factor; it can lead to serious damages. That is why the motion strategy should be optimized. We search the fastest sorting, with the minimum internal loads. The travelled distance during each pick and place cycle should be minimized, as well as the highest acceleration, the velocity, and specially the actuation torque. Thus, an important part of our study is to design an algorithm that minimizes cycle duration. The robot works in waste sorting. It has to be fast. It should also place each kind of waste in the specific container. An optimization process for the pick and place strategy, the order of the picked object, is essential.

In the following a literature review about the parallel mechanisms used in the pick and place field is presented in Sect. 2. The problem settings and the sorting stages are presented in Sect. 2. The trajectory design for the moving platform and the optimization algorithm are described in Sect. 4. The kinematics calculations of the delta robot, with the horary equations of the actuators are discussed in Sect. 5. To validate the algorithm, a small prototype is shown in Sect. 6. We finish with some conclusions and further works.

2 State of the Art

Pick and-place operations are required in most light industry sectors such as waste sorting, electronics, packaging, pharmacy and many others. Manipulated objects are light in weight and small in size, but in large amount. Since the charge is cumulative in a serial robot, the inertia becomes a serious load. The task load becomes negligible with respect to the inertia. That is why Clavel presented his invention of a 3 degree of freedom (DOF) translational parallel mechanism named Delta [2–6]: a fast parallel robot based on parallelogram mechanisms. Actuators are base-mounted, so they do not move. Low-mass links induce low global inertia. Thus the Delta robot is an ideal candidate for high-speed pick-and-place operations. The ratio between the link and parallelogram lengths governs the cylindrical workspace diameter/height ratio. The relevant literature along this line can be exemplified by recent publications [7–18] and many others.

Extensive research activities have been directed towards analysis of the motion strategy [19–26]. In fact, for the same displacement, the maximum velocity and acceleration can vary significantly upon the motion strategy, which leads to inertia forces from the loads on the actuators. Large inertia forces lead to the use of heavy control and less accuracy [27].

This paper deals with the optimization of the pick and place operation. We present an algorithm to maximize the pick and place process. The duration depends mainly on the cycle time and the intermediate interval between two cycles. The order of the picked objects is also important. The algorithm tries to minimize the total duration of two consecutive cycles.

3 Problem Settings

Our goal is to robotize the waste sorting. Thus, wastes are disposed on a conveyor. The first stage consists of an electro-magnet in order to retain all ferromagnetic wastes. In a second stage all non-ferromagnetic metallic parts are ejected from the conveyor using an eddy-current system. In the third stage, plastics and glass are to be sorted by the robot (Fig. 1). Finally, only organic wastes remain on the conveyor.

A special infrared camera can detect the plastics and the glass on the conveyor (Fig. 2). It gives also the y coordinate of the plastic or glass component.

The conveyor speed v_c is known and constant, which allows us to compute the x coordinate function of time. Therefore, the duration needed for the detected object to enter into the robot workspace can be calculated, as well as the duration inside the workspace.

In order to optimize the pick and place process, the objective is to maximize the number of pick-and place cycles per minute. The constraint to respect is to pick the object before getting off the robot workspace (Fig. 3). The positions $\{x_c, y_p, z_c\}$ and $\{x_c, y_g, z_c\}$ of the containers are used as inputs. Once an object passes through

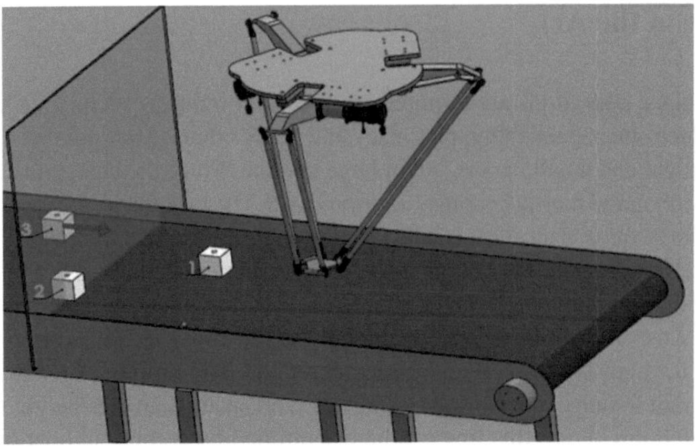

Fig. 1 Robotized waste sorting system

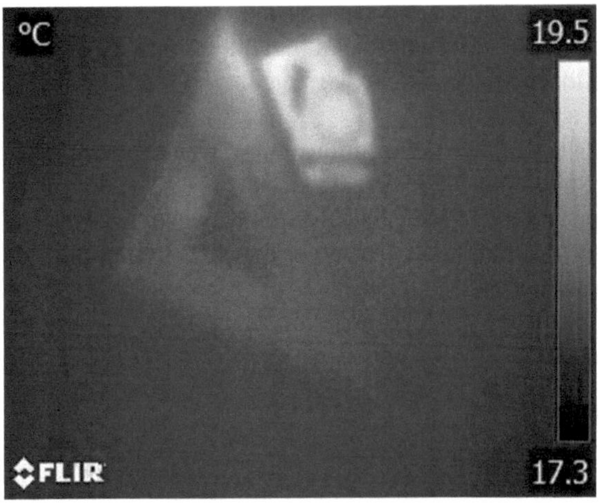

Fig. 2 Plastic and glass parts detected by Flir A35

the detector, its y coordinate and its corresponding container are defined. The x coordinate becomes

$$x = v_c.t. \tag{1}$$

To explain the idea, let us consider two objects. The first detected object must be picked from the target position p_{1t}, the second from the target position p_{2t}. If the detected object is plastic, the robot will place it at point p, otherwise at point g. To remove both objects, the moving platform has to carry out three travels, upon two choices:

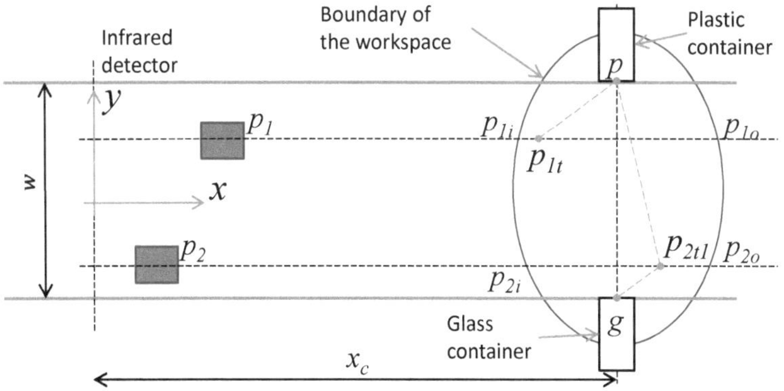

Fig. 3 Pick and place operation

- Travel 1 from p_{1t} to the corresponding container;
- Travel 2 from the previous container to p_{2t};
- Travel 3 from p_{2t} to the corresponding container.

The second choice is to pick the second object before the first one. In this case, travel 2 starts at the container and ends at p_{2t} (Fig. 3).

In both cases there are two pick and place travels and an intermediate travel. Then, the algorithm searches for the target points, p_{1t}, p_{2t}, that minimize the total duration of the three travels. A third binary variable ε is needed. Let ΔT_1 and ΔT_2 be the durations of the pick and place travels, ΔT_{i1} (ΔT_{i2}) the duration of the intermediate travel if object 1 (object 2) is picked first.

For better performance, it will be better to consider three detected objects instead of two. But in this case there are six possible choices. Finally, in order to respect the constraints, the algorithm uses the penalty method.

4 Trajectory Design and Optimization

The manipulator has to pick the part from the target point and place it in the container. Then, it has to lift up the part, move it until it reaches the container and finally place it down (Fig. 4). To facilitate the problem, the motion is assumed in a vertical plane, i.e. the motion along y is linear with the motion along x.

$$y = \frac{y_p - y_{p_t}}{x_p - x_{p_t}}(x - x_{p_t}) + y_{p_t}$$

$$\dot{y} = \frac{y_p - y_{p_t}}{x_p - x_{p_t}}\dot{x}. \qquad (2)$$

Fig. 4 Trajectory of the moving platform

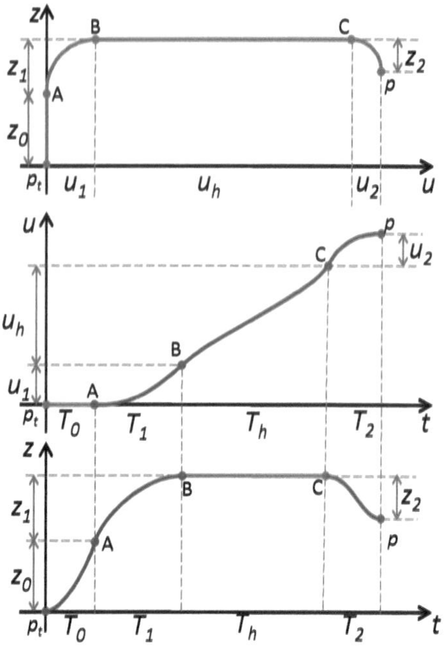

x_{p_t} and y_{p_t} are the coordinates of the target point where the object is picked. In order to keep only two variables a variable u such as:

$$\dot{\vec{u}} = \dot{x}\vec{i} + \dot{y}\vec{j}. \tag{3}$$

is considered. Consequently, the design of the motion along u and z is carried out. This variable u also helps to avoid the excess of the maximum velocity and acceleration.

The motion design has to give the shortest time for the cycle with full respect to the initial conditions. The motion is designed as a function of time along u and z separately.

4.1 Initial Conditions

The path is divided into four phases (Fig. 4). Phase *0* (p_tA) is a vertical line straight motion, phase h (BC) is a horizontal straight line, phases *1* (AB) and *2* (Cp) are blended. At A, B and C the continuity of velocity and acceleration is a must. The initial conditions to respect are:

Phase p_tA:

$$x_{p_tA} = 0 \quad z_{p_tA} = a_3 t_0^3 + a_2 t_0^2 \tag{4}$$

Phase AB:

$$\begin{matrix} \dot{u}_A \neq 0 \; \ddot{u}_A = 0 \\ \dot{u}_B \neq 0 \quad \forall \ddot{u}_B \end{matrix} \Rightarrow u_{AB} = U_1\left(\frac{t_0}{T_1} - \frac{1}{\pi}\sin\left(\frac{\pi t_0}{T_1}\right)\right) \tag{5}$$

$$\begin{matrix} \dot{z}_A \neq 0 \quad \forall \ddot{z}_A \\ \dot{z}_B \neq 0 \; \ddot{z}_B = 0 \end{matrix} \Rightarrow z_{AB} = Z_1\left(\frac{t_0}{T_1} + \frac{1}{\pi}\sin\left(\frac{\pi t_0}{T_1}\right)\right) + Z_0 \tag{6}$$

Phase BC:

$$u_{BC} = b_3 t_h^3 + b_2 t_h^2 + b_1 t_h + b_0 \; z_{BC} = Z_0 + Z_1 \tag{7}$$

Phase Cp:

$$\begin{matrix} \dot{u}_C \neq 0 \; \ddot{u}_C \, no\, matter \\ \dot{u}_D = 0 \; \ddot{u}_p \, no\, matter \end{matrix} \Rightarrow u_{Cp} = U_2 \sin\left(\frac{\pi t_2}{2T_2}\right) \tag{8}$$

$$\begin{matrix} \dot{z}_C = 0 \quad \ddot{z}_C = 0 \\ \dot{z}_D = 0 \; \ddot{z}_p \, no\, matter \end{matrix} \Rightarrow z_{Cp} = Z_2\left(1 - \frac{t_2}{T_2} + \frac{1}{2\pi}\sin\left(\frac{2\pi t_2}{T_2}\right)\right) + Z_0 + Z_1 \tag{9}$$

Known values:

$$U_1 + U_h + U_2 = u_p - u_{p_i}; \Delta Z = Z_0 + Z_1 + Z_2 \tag{10}$$

4.2 Motion Along x

$$u_B = \frac{2U_1}{T_1} = x_{BC}(t_h = 0) \Rightarrow b_0 = U_1 \tag{11}$$

$$\dot{u}_B \equiv \dot{u}_{AB}(t_1 = T_1) = \dot{u}_{BC}(t_h = 0) = \frac{2U_1}{T_1} = b_1 \tag{12}$$

$$\ddot{u}_B \equiv \ddot{u}_{AB}(t_1 = T_1) = \ddot{u}_{BC}(t_h = 0) = 0 \Rightarrow b_2 = 0 \tag{13}$$

$$\ddot{u}_C = 0 \Rightarrow b_3 = 0 \tag{14}$$

The motion during phase BC is uniform:

$$u_{BC} = 2U_1\frac{T_h}{T_1}\left(\frac{t_h}{T_h}\right) + U_1 \tag{15}$$

$$u_e = U_1 + U_h = u_{BC}(t_h = T_h) = 2U_1\frac{T_h}{T_1} + U_1$$

$$U_h = 2\frac{T_h}{T_1}U_1 \tag{16}$$

$$\dot{u}_C = \dot{u}_{BC}(t_h = T_h) = \dot{u}_{CD}(t_3 = 0) = \frac{2U_1}{T_1} = \frac{\pi U_2}{2T_2} \Rightarrow U_2 = 4\frac{T_2}{\pi T_1}U_1 \tag{17}$$

4.3 Motion Along z

$$Z_A = Z_0 = a_3 T_0^3 + a_2 T_0^2 \tag{18}$$

$$\dot{z}_A = \dot{z}_{OA}(t_0 = T_0) = \dot{z}_{AB}(t_1 = T_1) \Rightarrow 3a_3 T_0^2 + 2a_2 T_0 = \frac{2Z_1}{T_1} \tag{19}$$

$$z_{OA} = \frac{2}{T_0^2}\left(-\frac{Z_0}{T_0} + \frac{Z_1}{T_1}\right)t_0^3 + \left(\frac{3Z_0}{T_0} - \frac{2Z_1}{T_1}\right)\frac{t_0^2}{T_0} \tag{20}$$

$$\ddot{z}_A = \ddot{z}_{OA}(t_0 = T_0) = \ddot{z}_{AB}(t_1 = T_1) = 0 \Rightarrow Z_0 = \frac{10}{9}\frac{T_0}{T_1}Z_1 \tag{21}$$

$$z_{OA} = -\frac{Z_0}{5}\left(\frac{t_0}{T_0}\right)^3 + \frac{6}{5}Z_0\left(\frac{t_0}{T_0}\right)^2 \tag{22}$$

4.4 Velocity and Acceleration Conditions

At A, z is at the maximum, $\dot{u}$ is zero. At p_t, $\ddot{z}$ is at the maximum. Thus, we have to impose $\ddot{z}_{p_t} = \ddot{z}_{max}$ and $\dot{z}_A = \dot{z}_{max}$. At B, $\dot{u}$ is at the maximum and $\dot{z}$ is zero. Than we have to impose $\dot{u}_{BC} = \dot{u}_{max}$. At p, the acceleration along x, in magnitude, must be less then $\ddot{u}_{max} \cdot \dot{u}_{max}$, $\ddot{u}_{max}$, $\dot{z}_{max}$ and $\ddot{z}_{max}$ depend on the used actuators and on the robot structure.

$$\dot{u}_B = \dot{u}_{BC} = \dot{u}_B = \dot{u}_{max} = \frac{2U_1}{T_1} = \frac{\pi U_2}{2T_2} \tag{23}$$

$$\dot{z}_A = \dot{z}_{max} = \frac{2Z_1}{T_1} \tag{24}$$

$$\ddot{z}_{p_t} = \ddot{z}_{max} = \frac{12}{5}\frac{Z_0}{T_0^2} \tag{25}$$

$$\ddot{u}_p = -\frac{\pi^2 U_2}{4T_2^2} = -\ddot{u}_{max} \tag{26}$$

Equations (10), (21), (24) and (25) give:

$$\begin{aligned} T_0 &= \tfrac{4}{3}\tfrac{\dot{Z}_{max}}{\ddot{Z}_{max}} & Z_0 &= \tfrac{20}{27}\tfrac{\dot{Z}_{max}^2}{\ddot{Z}_{max}} \\ T_1 &= 2\tfrac{\Delta Z}{\dot{Z}_{max}} - \tfrac{40}{27}\tfrac{\dot{Z}_{max}}{\ddot{Z}_{max}} & Z_1 &= \Delta Z - \tfrac{20}{27}\tfrac{\dot{Z}_{max}^2}{\ddot{Z}_{max}} \end{aligned} \tag{27}$$

Equations (10), (16), (13) and (26) give

$$U_1 = \frac{\dot{x}_{max} T_1}{2} \qquad U_2 = \pi \frac{\dot{x}_{max}^2}{\ddot{x}_{max}} \qquad U_h = \Delta U - U_1 - U_2$$
$$T_2 = \frac{\pi}{2} \frac{\dot{x}_{max}}{\ddot{x}_{max}} \; T_h = \frac{\Delta X}{\dot{x}_{max}} - \pi \frac{\dot{x}_{max}}{\ddot{x}_{max}} - \frac{T_1}{2} \tag{28}$$

Finally the total time of the pick and place operation is:

$$\Delta T = \frac{16}{27} \frac{\dot{z}_{max}}{\ddot{z}_{max}} + \left(\frac{\pi}{2} - 1\right) \frac{\dot{u}_{max}}{\ddot{u}_{max}} + \frac{\Delta Z}{\dot{z}_{maz}} + \frac{\Delta U}{\dot{u}_{max}} \tag{29}$$

The objective function now is:

$$F = \Delta T_1 + \Delta T_2 + \varepsilon \Delta T_{i1} + \bar{\varepsilon} \Delta T_{i2}$$
$$\varepsilon = \begin{cases} 1 \text{ if the first object is picked before} \\ 0 \text{ if the second object is picked before} \end{cases} \tag{30}$$

The optimization variables are $x_{p_{1t}}$ and $x_{p_{2t}}$, the abscissas of the target points p_{t1} and p_{t2} where objects are picked, and ε is the binary variable that defines the picking order.

F should be minimized under the constraint, both objects must be inside the workspace at the pick instant i.e.:

$$\frac{x_{p_{1i}}}{x_{p_{t1}}} - 1 < 0 \; \frac{x_{p_{2i}}}{x_{p_{t2}}} - 1 < 0$$
$$\frac{x_{p_{t1}}}{x_{p_{1o}}} - 1 < 0 \; \frac{x_{p_{t2}}}{x_{p_{2o}}} - 1 < 0 \tag{31}$$

In order to take Eq. (31) into consideration, the penalty method is used. The penalized objective function is:

$$L = F + \lambda_1.sup\left(\left(\frac{x_{p_{1i}}}{x_{p_{t1}}} - 1\right), 0\right) + \lambda_2.sup\left(\left(\frac{x_{p_{2i}}}{x_{p_{t2}}} - 1\right), 0\right)$$
$$+ \lambda_3.sup\left(\left(\frac{x_{p_{t1}}}{x_{p_{1o}}} - 1\right), 0\right) + \lambda_4.sup\left(\left(\frac{x_{p_{t2}}}{x_{p_{2o}}} - 1\right), 0\right) \tag{32}$$

$\lambda_1, \lambda_2, \lambda_3, \lambda_4$ are penalty constant coefficients

The algorithm becomes (Fig. 5).

5 Kinematics

The motion calculated is for the moving platform. However, the controller acts on the actuators. Thus, the equations of the joint coordinate motions must be calculated. In this section the inverse kinematics model is derived. Then it will be sufficient to replace x, y and z by their values to get θ_1, θ_2 and θ_3 as functions of time. The Delta

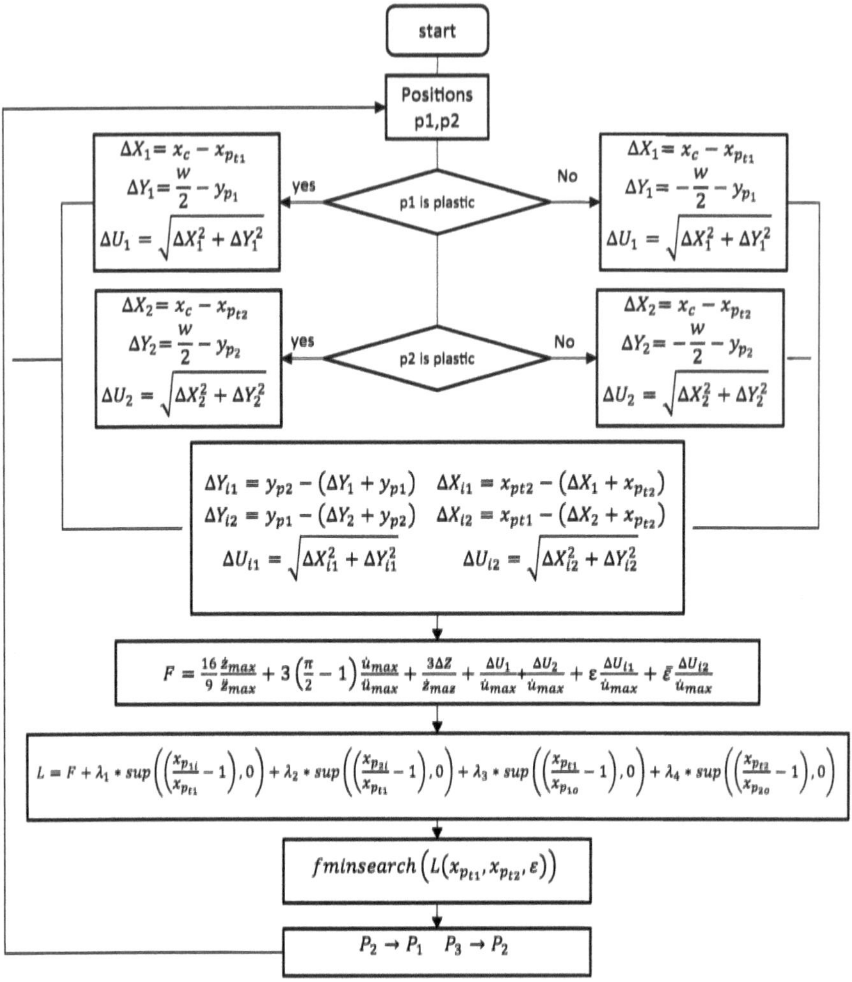

Fig. 5 Pick and place process optimization algorithm

robot has three degrees of freedom (dof), thus it can carry out three translations for its moving platform in the workspace. It is the assembly of three kinematic chains where each one can be considered as a leg. The hip centers are the points B_i, the knee centers are A_i, and the ankle centers are P_i, for leg i. Hips are three revolute joints, thus the fixed platform will be considered as an equilateral triangle made by the tree joint axes. The length of this triangle side is denoted by s_B. The moving platform is also an equilateral triangle but obtained by its three edges P_i. The length of this triangle side is denoted by s_p (Fig. 6). The motion of the moving platform is only a translation, thus it has a constant orientation.

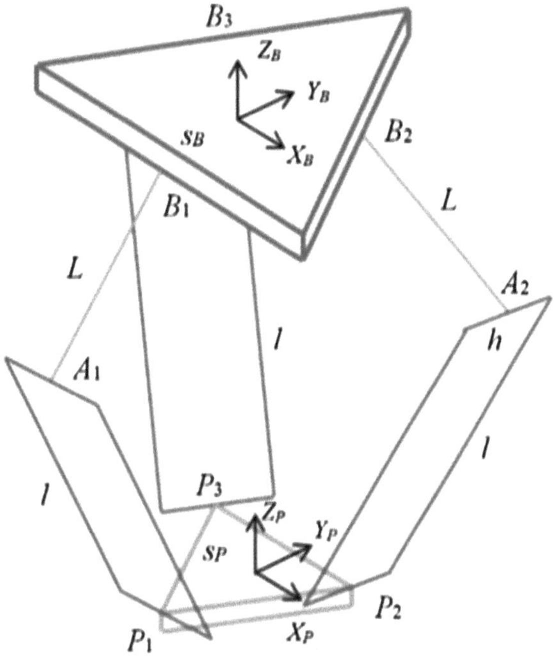

Fig. 6 Delta robot kinematic diagram

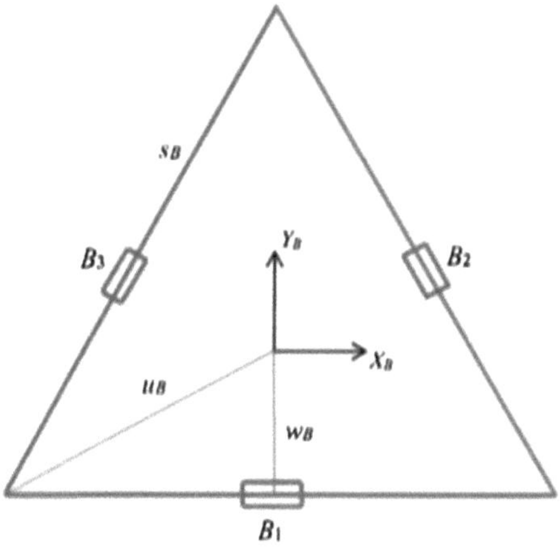

Fig. 7 Delta robot fixed base details

We associate the reference frame $\{B\}$ to the fixed platform with the origin at the base center. We associate the reference frame $\{P\}$ to the moving platform, with the origin at the center of the triangle $P_1P_2P_3$. The rotation matrix ${}^B_P R = I_3$ remains constant. The joint coordinates are $\{\theta_1, \theta_2, \theta_3\}^{\mathrm{T}}$ and the operational coordinates are ${}^B P_P = \{xyz\}^T$. The design shown in Figs. 7 and 8 presents the symmetry of the robot architecture.

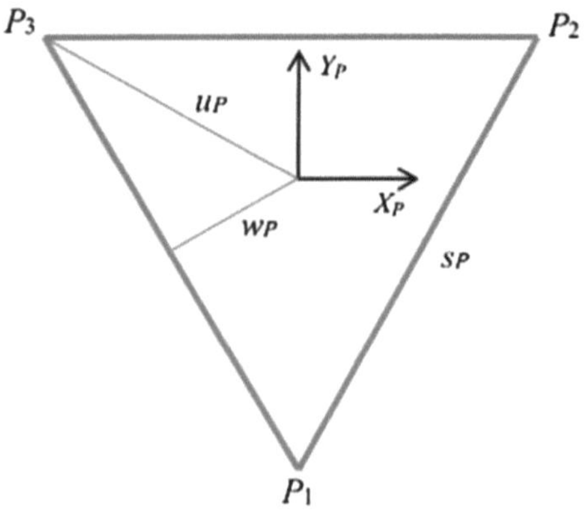

Fig. 8 Delta robot moving platform details

The hip B_i coordinates are constant in the coordinate system $\{B\}$ and the ankle P_i coordinates are constant in the base frame $\{P\}$:

$$^{B}B_1 = \begin{Bmatrix} 0 \\ -w_B \\ 0 \end{Bmatrix} \; ^{B}B_2 = \begin{Bmatrix} \frac{\sqrt{3}}{2}w_B \\ \frac{1}{2}w_B \\ 0 \end{Bmatrix} \; ^{B}B_3 = \begin{Bmatrix} -\frac{\sqrt{3}}{2}w_B \\ \frac{1}{2}w_B \\ 0 \end{Bmatrix}$$

$$^{P}P_1 = \begin{Bmatrix} 0 \\ -u_P \\ 0 \end{Bmatrix} \; ^{P}P_2 = \begin{Bmatrix} \frac{s_P}{2} \\ w_P \\ 0 \end{Bmatrix} \; ^{P}P_3 = \begin{Bmatrix} -\frac{s_P}{2} \\ w_P \\ 0 \end{Bmatrix} \tag{33}$$

$$w_B = \frac{\sqrt{3}}{6}s_B \; u_B = \frac{\sqrt{3}}{3}s_B$$
$$w_P = \frac{\sqrt{3}}{6}s_P \; u_P = \frac{\sqrt{3}}{3}s_P$$

N.B.: The upper left index is the base where the vector is described.

Since three kinematics chains exist (Fig. 6), three vector-loop closure [28] equations should be elaborated:

$$\{^{B}B_i\} + \{^{B}L_i\} + \{^{B}l_i\} = \{^{B}P_P\} + [^{B}_{P}R]\{^{P}P_i\} \tag{34}$$

For $i = 1, 2, 3$.

The operational coordinates are $^{B}P_P = \{xyz\}^T$. Vectors $\{^{B}L_i\}$ depend on the joint coordinates:

$$
{}^{B}L_1 = \begin{Bmatrix} 0 \\ -L\,\cos\theta_1 \\ -L\,\sin\theta_1 \end{Bmatrix}
$$

$$
{}^{B}L_2 = \begin{Bmatrix} \frac{\sqrt{3}}{2}L\,\cos\theta_2 \\ \frac{1}{2}L\,\cos\theta_2 \\ -L\,\sin\theta_2 \end{Bmatrix}
$$

$$
{}^{B}L_3 = \begin{Bmatrix} -\frac{\sqrt{3}}{2}L\,\cos\theta_3 \\ \frac{1}{2}L\,\cos\theta_3 \\ -L\,\sin\theta_3 \end{Bmatrix}
$$

$$
\left\{{}^{B}l_1\right\} = \begin{Bmatrix} x \\ y + L\,\cos\theta_1 + a \\ z + L\,\sin\theta_1 \end{Bmatrix}
$$

$$
\left\{{}^{B}l_2\right\} = \begin{Bmatrix} x - \frac{\sqrt{3}}{2}L\,\cos\theta_2 + b \\ y - \frac{1}{2}L\,\cos\theta_2 + c \\ z + L\,\sin\theta_2 \end{Bmatrix}
$$

$$
\left\{{}^{B}l_3\right\} = \begin{Bmatrix} x + \frac{\sqrt{3}}{2}L\,\cos\theta_3 - b \\ y - \frac{1}{2}L\,\cos\theta_3 + c \\ z + L\,\sin\theta_3 \end{Bmatrix} \tag{35}
$$

where: $a = w_b - u_P, b = \frac{s_P}{2} - \frac{\sqrt{3}}{2}w_B, c = w_p - \frac{1}{2}w_B$

Equation (34) establish respectively the kinematics equations for the mechanism:

$$
2L(y+a)\cos\theta_1 + 2zL\,\sin\theta_1 + x^2 + y^2 + z^2 + a^2 + L^2 + 2ya - l^2 = 0 \tag{36}
$$

$$
-L\left(\sqrt{3}(x+b) + y + c\right)\cos\theta_2 + 2zL\,\sin\theta_2 + x^2 + y^2 + z^2 + b^2 + c^2 + L^2 + 2xb + 2yc - l^2 = 0 \tag{37}
$$

$$
L\left(\sqrt{3}(x-b) - y - c\right)\cos\theta_3 + 2zL\,\sin\theta_3 + x^2 + y^2 + z^2 + b^2 + c^2 + L^2 - 2xb + 2yc - l^2 = 0 \tag{38}
$$

The three independent scalar equations are of the form:

$$
E_i\cos\theta_i + F_i\sin\theta_i + G_i = 0 \quad i = 1, 2, 3 \tag{39}
$$

where E_i, F_i, G_i are dependent of the coordinates $\{x, y, z\}$. Using the tangent half-angle substitution:

$$t_{l_{1,2}} = \frac{-F_l \pm \sqrt{E_l^2 + F_l^2 - G_l^2}}{G_l - E_l} \tag{40}$$

Then:

$$\theta_l = 2\tan^{-1}(t_l) \tag{41}$$

Two inverse kinematic model solutions are possible for each leg of the Delta Robot. The first one is for a knee kinked out whereas for the second one, the knee is kinked in. In total, eight different combinations are possible. We will use the case where the three knees are kinked out.

The differential kinematics model can be obtained simply by differentiating Eq. (34) with respect to the time. Written in matrix-vector form:

$$[\boldsymbol{A}]\{\dot{\boldsymbol{X}}\} = [\boldsymbol{B}]\{\dot{\boldsymbol{\theta}}\} \begin{bmatrix} x & y + a + LC_1 & z + LS_1 \\ 2(x+b) - \sqrt{3}LC_2 & 2(y+c) - LC_2 & 2(z + LS_2) \\ 2(x-b) + \sqrt{3}LC_3 & 2(y+c) - LC_3 & 2(z + LS_3) \end{bmatrix} \begin{Bmatrix} \dot{x} \\ \dot{y} \\ \dot{z} \end{Bmatrix}$$
$$= \begin{bmatrix} b_{11} & 0 & 0 \\ 0 & b_{22} & 0 \\ 0 & 0 & b_{33} \end{bmatrix} \begin{Bmatrix} \dot{\theta}_1 \\ \dot{\theta}_2 \\ \dot{\theta}_3 \end{Bmatrix} \tag{42}$$

$$\begin{aligned} b_{11} &= L[(y+a)S_1 - zC_1] \\ b_{22} &= -L\left[\left(\sqrt{3}(x+b) + y + c\right)S_2 + 2zC_2\right] \\ b_{33} &= L\left[\left(\sqrt{3}(x-b) - y - c\right)S_3 - 2zC_3\right] \\ S_l &= \sin\theta_l \quad C_l = \cos\theta_l \end{aligned}$$

Substituting x, y and z in Eq. (41), gives the equations of the joint coordinates as function of time. Similarly, the substitution of $\dot{x}$, $\dot{y}$ and $\dot{z}$ in Eq. (42), yields a system of differential equations. This system can be solved numerically.

For an optimal control, a motion study has to be carried out showing multiple results concerning the torque of the actuators. The robot will work with a maximum payload of 2 kg. The dynamics model needs the solution of 54 equations with 54 unknowns. If the needed torque of at least an actuator exceeds its nominal torque, a correction is needed because the desired motion will be lost. In this case, the maximum velocities and accelerations, $\dot{u}_{max}$, $\ddot{u}_{max}$ $\dot{z}_{max}$ and $\ddot{z}_{max}$ should be moderated.

Fig. 9 Prototype

Fig. 10 Inductive detectors in the prototype

6 Prototype

In order to validate the algorithm, a small prototype is made using 3D printing technology. It is controlled using an Arduino-uno controller, and the code is written in Labview (Fig. 9).

The conveyor is animated with a velocity $v_c = 1$ m/s. The conveyor width is 12 cm. It is not possible to use the infrared detector with the prototype. For this reason, a series of inductive sensors and a series of capacitive sensors are used (Fig. 10).

Instead of glass and plastic, metallic and plastic elements are used. The plastic parts are detected by the capacitive sensor only and the metallic by both sensors. Thus, metallic elements can be separated from the plastic ones on the conveyor. The position of the sensor gives the y-coordinate. The previous algorithm can be used in the same way. And the prototype could sort up to 3 parts per second.

7 Conclusions and Further Works

In this paper we presented the integration of a parallel manipulator in a waste sorting system. Metallic objects are removed using electromagnets and eddy current. Plastic and glass are detected using an infrared sensor. Then the y coordinate is determined as well as the x coordinate of the object function of time. We calculated the trajectory and the equation of motion for the pick and place operation. We chose blending phases that ensure the continuity of acceleration and velocity. In order to get the fastest motion, the maximum possible values of the velocity and acceleration are targeted. We developed an algorithm to find the position where each object must be picked and in which order in the purpose of getting the maximum number of cycles per minute. In order to ensure that all objects are removed, non-equality constraints are included in the operational research. To take these constraints into consideration, the algorithm uses the penalty method: the objective function is penalized when a constraint is violated. Since the controller acts on the joint coordinates, we elaborated the inverse kinematics and differential models. Those models give the equation of the joints motion as function of time.

In this paper, the trajectory is considered in a vertical plane. The task can be performed much faster if a curved trajectory is used with conditioned velocity and acceleration at intermediate points. In this case, interesting effects will be found for the motion strategies. Moreover, our prototype is very light. Therefore, no automatic control was needed. However, on the real device the inertia of the robot is not negligible. The objective is to sort more than 200 tons of waste per day. Thus, we need very high velocities leading to high non-linear effects and disturbances. The control will be a complicated task and a classical PID controller will not work. Those aspects are going to be our future step.

References

1. Merlet JP (2000) Parallel robots. Kluwer Academic, Dordrecht
2. Bonev I (2001) Delta parallel robot—the story of success. http://www.parallelmic.org
3. Cervantez-Sanchez JJ, Rendon-Sanchez JG (1999) A simplified approach for obtaining the workspace of a class of 2-DOF planar parallel robot. Mech Mach Theory 34(7):1057–1073
4. Clavel R (1990) Device for the movement and positioning of an element in space. U.S. Patent 4976582, 11 Dec
5. Rey L, Clavel R (1999) The delta parallel robot. In: Boer CR, Molinari-Tosatti L, Smith KS (eds) Parallel kinematic machine. Springer, New York
6. Clavel R (1988) Delta, a fast robot with parallel geometry. In: The 18th international symposium on industrial robots (ISIR), Sydney, Australia, pp 91–100

7. Codourey A (1996) Dynamic modeling and mass matrix evaluation of the delta parallel robot for axes decoupling control. In: Proceedings of IEEE/RSJ international conference on intelligent robots and systems, pp 1211–1218
8. Maurine P, Dombre E (1996) A calibration procedure for the parallel robot Delta 4. In: Proceedings of IEEE international conference on robotics and automation (ICRA'96), Minneapolis, MN, pp 975–980
9. Miller K (1992) The proposal of a new model of direct drive robot Delta-4 dynamics. In: Proceedings of international conference on advanced robotics, Tokyo, Japan, pp 411–416
10. Miller K (1995) Experimental verification of modeling of Delta robot dynamics by direct application of Hamilton's principle. In: Proceedings of IEEE international conference on robotics and automation (ICRA'95), Nagoya, Japan, pp 532–537
11. Miller K (1995) On accuracy and computational efficiency of Delta direct drive robot dynamics mode. In: Proceedings of international symposium on microsystems, intelligent materials and robots, 1995, Sendai, Japan, pp 568–571
12. Miller K (1995) Model-based control of Delta direct drive parallel robot; Trajectory tracking experiments. In: Proceedings of the 26th international symposium of industrial robots, Singapore, pp 491–496
13. Miller K (1995) Modeling of dynamics and model-based control of Delta direct-drive parallel robot. J Robot Mechatron 17(4):344–352
14. Miller K, Clavel R (1992) The lagrange-based model of Delta-4 robot dynamics. Robot Sys 8(4):49–54
15. Stamper RC, Tsai LW, Walsh GC (1997) Optimization of a 3-DOF translational platform for well-conditioned workspace. In: Proceedings Of IEEE international conference on robotics and automation (ICRA'97), Albuquerque, pp 3250–3255
16. Tsai LW, Walsh GC, Stamper RC (1996) Kinematics of a novel 3-DOF translational platform. In: Proceedings of IEEE international conference on robotics and automation (ICRA'96), Minneapolis, MN, pp 3446–3451
17. Tsai LW (1999) The enumeration of a class of 3-DOF parallel manipulators. In: Proceedings of the 10th world congress on the theory of machine and mechanisms, Oulu, Finland, pp 1121–1126
18. Tsai LW, Joshi S (2002) Kinematics and optimization of a 3UPU parallel manipulator. ASME J Mech Des 4(122):439–446
19. Bobrow JE, Dubowsky S, Gibson JS (1985) Time-optimal control of robotic manipulators along specified paths. Int J Robot Res 4(3):3–17
20. Braibant V, Geradin M (1987) Optimum path planning of robot arms. Robotica 5:323–331
21. Dubowsky S, Shiller Z (1984) Optimal dynamic trajectories for robotic manipulators. In: Proceedings of the fifth CISMIFToMM symposium on theory and practice of robots and manipulators, pp 133–143
22. Kim BK, Shin KG (1985) Minimum-time path planning for robot arms and their dynamics. IEEE Trans Sys Man Cybern SMC 15(2):213–223
23. Lin CS, Chang PR. Luh JYS (1983) Formulation and optimization of cubic polynomial joint trajectories for industrial robots. IEEE Trans Autom Control AC 28(12):1066–1074
24. Lin CS, Chang PR (1985) Approximate optimum paths of robot manipulators under realistic physical constraints. In: Proceedings of the IEEE international conference on robotics and automation, New York, pp 737–742
25. Luh JYS, Lin CS (1981) Optimal path planning for mechanical manipulators. ASME J Dyn Sys Meas Control 102(6):142–1.51
26. Shiller Z, Dubowsky S (1985) On the optimal control of robotic manipulators with actuator and end effector constraints. In: Proceedings of the IEEE conference on robotics and automation, New York, pp 614–620
27. Pateloup V, Duc E, Ray P (2010) B-spline approximation of circle arc and straight line for pocket machining. Comput Aided Des 42(9):817–827
28. Briot S, Khalil W (2015) Dynamics of parallel robots. In: From rigid bodies to flexible elements. Springer

Creative Design of 2-DOF, 7-Link Multiloop Robotic Mechanisms

Vladimir Pozhbelko and Ekaterina Kuts

Abstract Structural synthesis of kinematic chains and multiple DOF multiloop mechanisms with specified input parameters is the first and the most important stage in the mechanical conceptual creative design of robots. This paper proposes a new approach to structural synthesis and creative design of closed-loop robotic mechanism with simple and multiple joints based on an original addition synthesis method. First, the structural synthesis method of non-fractionated and fractionated multiple joint kinematic chains with the smallest closed loop is presented. Then, the complete atlas database of all feasible 12 types of 2-DOF, 7-link, 2-basic (independent) loops generalized simple and multiple joint kinematic chain and corresponding atlas of hinged mechanisms are obtained. Third, based on classified atlas databases of kinematic chains and mechanisms with simple and multiple joints, a creative design method is proposed to obtain all the feasible mechanisms for a specified task. Finally, examples of the creative design of single arm robot manipulators are also conducted to demonstrate the effectiveness of the proposed method.

Keywords Structural synthesis · Multi-degree-of-freedom mechanisms · Creative design · Robot manipulators

Nomenclature

$\tilde{n}$ total number of links in the kinematic chains:

$$\tilde{n} = n_1 + n_2 + n_3 + \cdots + n_i;$$

n_1 the number of singular links;

V. Pozhbelko · E. Kuts (✉)
South Ural State University (National Research University), Chelyabinsk, Russia
e-mail: ekaterina.n.kuts@gmail.com

V. Pozhbelko
e-mail: pozhbelkovi@susu.ru

R. Rizk and M. Awad (eds.), *Mechanism, Machine, Robotics and Mechatronics Sciences*, Mechanisms and Machine Science 58,
https://doi.org/10.1007/978-3-319-89911-4_14

n_2	the number of binary links;
n_3	the number of ternary links;
n_i	the number of links having i joints with other links in the kinematic chain;
ν_j	the number of multiple joints;
ν_2	the number of two-pin joints (denoted by sign "j_2");
ν_3	the number of three-pin joints (denoted by sign "j_3");
V	resulted number of multiple joints ($V = \Sigma(j-1)\nu_j$);
V	total multiple joint factor, $V = 0$: simple joint kinematic chains, $V > 0$: multiple joint kinematic chains;
$V_{\max}$	the max value of resulted number of multiple joints;
K	the number of independent closed loops in the kinematic chains;
F	the number of degrees of freedom of mechanical system with the fixed link (DOF = F);
L_i	independent closed loop;
L_0	the number of sides in the peripheral loop in the kinematic chain;
$[L.A.]$	link assortment in the kinematic chain;
$[M.J.A.]$	multiple joint assortment in the kinematic chain;
$[L_\alpha]$	loop assortment in the kinematic chain.

1 Introduction

Structural synthesis of all possible non-isomorphic types of kinematic chains and construction of the complete atlases of all feasible mechanisms with given input parameters is one of the important and complex topological problems in the field of the conceptual design stage of mechanical engineering [1–23].

Until now, the most often studied planar hinged kinematic chains in structural synthesis are kinematic chains with simple joints. Note that, beside simple joint kinematic chains the various mechanical hinged systems can include multiple joint kinematic chains, that provides more simple construction and minimizes the space requirement. It is especially important in design of robots. However, structural synthesis of multibody mechanisms with multiple joint kinematic chains is difficult [24–31].

It is well known, that multiloop and multiple DOF (degrees of freedom) mechanisms have an extended diapason of output characteristics and therefore are widely used in various branches of engineering machinery, such as robotics machine equipment, high precision tools and microelectromechanical systems [12–17, 21, 23].

However, only part of the feasible mechanisms which satisfy a specified design task can be conceived on the designer's experience and intuition. Therefore, developing a more effective synthesis methods for synthesis of all the possible types of multiloop kinematic chains with simple and multiple joint is a most difficult and important problem in mechanism and machine theory up to present time.

The main purpose of the paper is to present a new a approach to structural synthesis and creative design of simple and multiple joint robotic mechanisms on the basis of original addition synthesis method for construction of complete atlas of all feasible 2-DOF, 7-link, 2-basic independent loops topological structures and creation of robot manipulators.

2 Terminology and Basic Definitions

The following terminology and definitions are needed and fundamental for the discussion presented in this paper.

2.1 Link Assortment

The type of links (by number of i joints incident to the link) and their number in the kinematic chain that are necessary to form the kinematic chain with given input parameters (for example, F, K and V) are defined as the "*link assortment*".

The link assortment array of closed kinematic chain is expressed as:

$$[L.A.] = [n_2, n_3, n_4, \ldots, n_i] \tag{1}$$

where n_2, n_3, n_4, …, n_i are the number of binary, ternary, quaternary, …, i-nary links, etc., in order, having i joints incident to the link.

2.2 Multiple Joint Assortment

The type of multiple joints (by number of $j \geq 2$ joint multiplicity in kinematic chains with number of basic loops $K \geq 2$ only) and their number (ν_2, ν_3, ν_4, …, ν_j) in open, closed and mixed kinematic chains is represented by [30]:

$$[M.J.A.] = \left[\nu_2 \bullet \nu_3 \bullet \nu_4 \bullet \ldots \bullet \nu_j\right] \tag{2}$$

where ν_2, ν_3, ν_4, …, ν_j are the number of two-pin joints (it is denoted as j_2 in figures), there-pin joints (it is denoted as j_3 in figures), four-pin joints (it is denoted as j_4 in figures), …, j-pin joints, etc., respectively.

It should be noted that in kinematic chains with simple joints only $[M.J.A.] = [0]$ takes place and all the possible sets of $[M.J.A.]$ for various K-loops multiple jointed kinematic chains were exactly calculated for the first time in the work [30].

2.3 Loop Assortment

The orderly of sequence of all closed loops in the kinematic chain including the peripheral loop L_0 (where type of loop is defined as the number of links participating on the periphery of any loop $L = K + 1$) is represented by [31]:

$$[L_\alpha] = [\alpha_3{-}\alpha_4{-}\alpha_5{-} \cdots {-}\alpha_m] = [3{-}4{-}5{-} \cdots {-}m] \quad (3)$$

where $\alpha_3, \alpha_4, \alpha_5, \ldots, \alpha_m$ indicate the number of links participating on the periphery of any loop for any simple and multiple joint kinematic chains. Note that α_3 denotes the smallest closed loop in the kinematic chain. Moreover, the following equality $K = L - 1$ takes place for any closed and mixed simple-jointed and multiple-jointed kinematic chains with total number L loops consisting of one peripheral loop (L_0) and others (located inside loop L_0) K-independent loops.

3 Directed Structural Synthesis of 2-DOF, 7-Link, 2-Basic Loops Kinematic Chains with Simple and Multiple Joints

Presented in this paper directed structural synthesis of all feasible kinematic chains with simple and multiple joints is based on addition synthesis method. This method uses different sets of various numbers of links (N_1, N_2) in closed (N_1) and in mixed or open (N_2) kinematic chains, where $N_1 + N_2 = \tilde{n} = const$ (in our case $\tilde{n} = 7 = const$).

The algorithm for directed synthesis of 2-DOF, 7-link, 2-basic loops generalized planar simple-jointed kinematic chains with the smallest closed loop $\alpha_{\min} = \alpha_3$ is shown below in Fig. 1 and the main steps are described as follows:

Step 1 Add 4-link mixed kinematic chain to ternary link of the 3-link closed kinematic chain consisting of 2 adjacent binary links and one 1 ternary link. As a result of structural synthesis, obtain the 7-link fractionated simple joint kinematic chain with [L.A.] = [5, 2, 0], $[L_\alpha] = [3 - 3 - 10]$, $L_0 = 10$ and one inversion of kinematic chain with the peripheral loop $L_0 = 3$.

Step 2 Add 4-link mixed kinematic chain to ternary link of the 3-link closed kinematic chain consisting of 2 adjacent binary links and one 1 ternary link. As a result, synthesize the 7-link fractionated simple joint kinematic chain with [L.A.] = [5, 2, 0], $[L_\alpha] = [3 -4 - 9]$, $L_0 = 9$ and two inversions of kinematic chain with the peripheral loops $L_0 = 4$ and $L_0 = 3$.

Step 3 Add 4-link open kinematic chain to 2 adjacent vertexes of quaternary link of the 3-link closed kinematic chain consisting of 2 adjacent binary links and 1 quaternary link. As a result of structural synthesis, obtain the 7-link fractionated simple joint kinematic chain with [L.A.] = [6, 0, 1], $[L_\alpha] = [3 - 5 -8]$, $L_0 = 8$ and two inversions of kinematic chain with the peripheral loops $L_0 = 5$ and $L_0 = 3$.

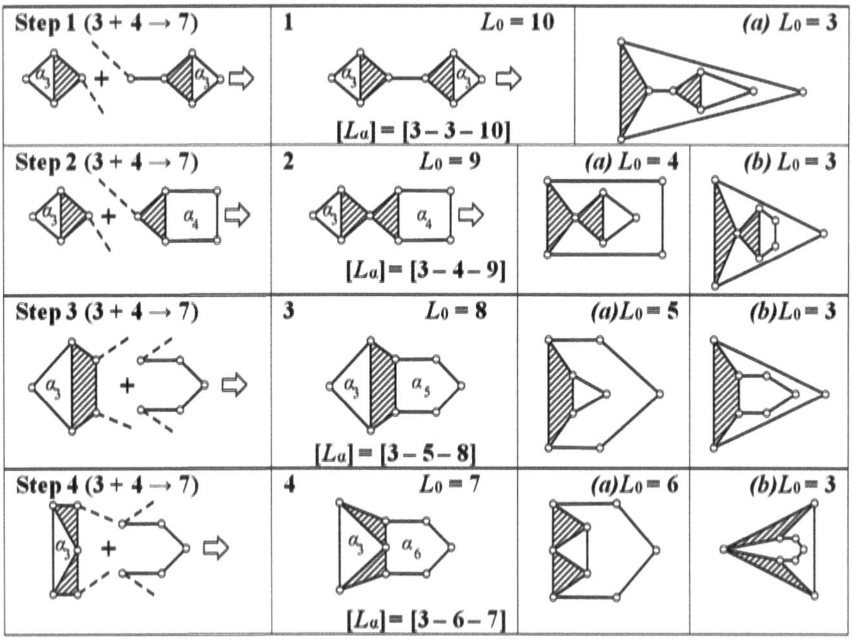

Fig. 1 The process to form and the atlas of all feasible 4 types of 2-DOF, 7-link, 2-basic loops generalized planar simple joint kinematic chains with the smallest close loop $\alpha_{\min} = \alpha_3$

Step 4 Add 4-link open kinematic chain to 2 adjacent ternary links of the 3-link closed kinematic chain consisting of 1 binary link and 2 adjacent ternary links. As a result, synthesize the 7-link non-fractionated simple joint kinematic chain with $[L.A.] = [5, 2, 0]$, $[L_\alpha] = [3 - 6 - 7]$, $L_0 = 7$ and two inversions of kinematic chain with the peripheral loops $L_0 = 6$ and $L_0 = 3$

In a similar manner, the algorithm for directed synthesis of 2-DOF, 7-link, 2-basic loops generalized planar multiple-jointed kinematic chains with the smallest closed loop $\alpha_{\min} = \alpha_3$ is shown below in Fig. 2 and the main steps are described as follows:

Step 1 Add 4-link closed kinematic chain to vertex of the 3-link closed kinematic chain consisting of 3 adjacent binary links. As a result of structural synthesis, obtain the 7-link multiple joint fractionated kinematic chains:

(1) with 1 multiple joint (j_2), (kinematic chain with closed loop α_3 (consisted of 3 binary links) and closed loop α_4 (consisted of 3 binary links and 1 ternary link) $[L.A.] = [6, 1, 0]$, $[L_\alpha] = [3 - 4 - 8]$ with $L_0 = 8$ $(V = 1)$);
(2) with 1 multiple joint (j_2), (kinematic chain with closed loop α_3 (consisted of 2 binary links and 1 ternary) and closed loop α_4 (consisted of 4 binary links) $[L.A.] = [6, 1, 0]$, $[L_\alpha] = [3 - 4 - 8]$ with $L_0 = 8$ $(V = 1)$);

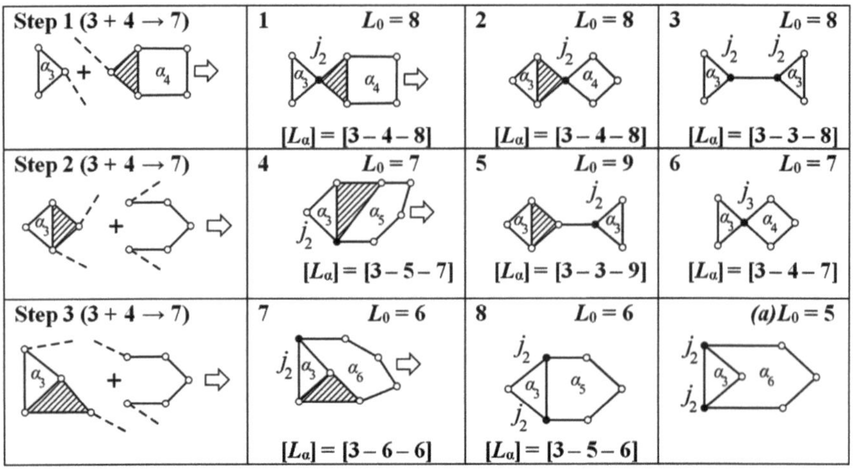

Fig. 2 The process to form and atlas of all feasible 8 types of 2-DOF, 7-link, 2-basic loop generalized planar multiple joint kinematic chains with the smallest close loop $\alpha_{\min} = \alpha_3$

(3) with 2 multiple joints (j_2, j_2), [$L.A.$] = [7, 0, 0], [L_α] = [3 – 3 – 8], L_0 = 8 (V = 2).

Step 2 Add 4-link open kinematic chain to ternary link and adjacent vertex of the 3-link closed kinematic chain consisting of 2 adjacent binary links and 1 ternary link. As a result, synthesize the 7-link multiple joint kinematic chains:

(4) with 1 multiple joint (j_2), [$L.A.$] = [6, 1, 0], [L_α] = [3 – 5 – 7], L_0 = 7 (V = 1);
(5) with 1 multiple joint (j_2), [$L.A.$] = [6, 1, 0], [L_α] = [3 – 3 – 9], L_0 = 9 (V = 1);
(6) with 1 multiple joint (j_3), [$L.A.$] = [7, 0, 0], [L_α] = [3 – 4 – 7], L_0 = 7 (V = 2).

Step 3 Add 4-link open kinematic chain to ternary link and non-adjacent vertex of the 3-link closed kinematic chain consisting of 2 adjacent binary links and 1 ternary link. As a result of structural synthesis, obtain the 7-link multiple joint non-fractionated kinematic chains:

(7) with 1 multiple joint (j_2), [$L.A.$] = [6, 1, 0], [L_α] = [3 – 6 – 6], L_0 = 6 (V = 1);
(8) with 2 multiple joints (j_2, j_2), two inversions (kinematic chain with [$L.A.$] = [7, 0, 0], [L_α] = [3 – 5 – 6] with L_0 = 6 (V = 2) and accordingly kinematic chain with [$L.A.$] = [7, 0, 0], [L_α] = [3 – 5 – 6]) with L_0 = 5 (V = 2).

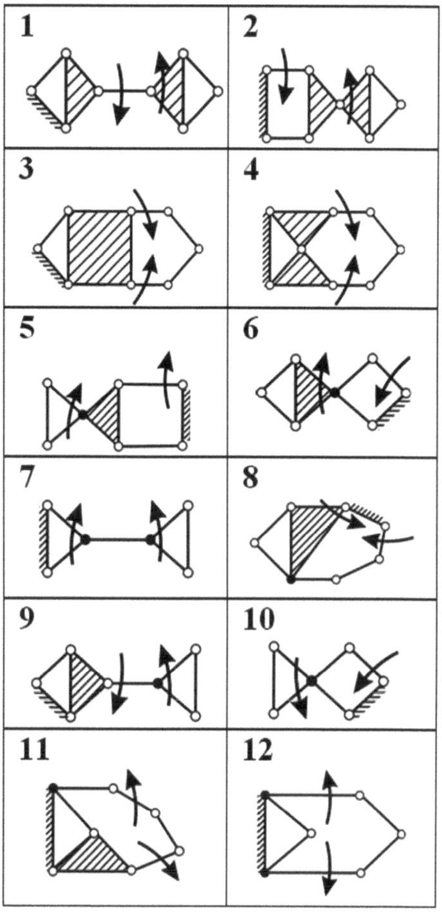

Fig. 3 Atlas of all 12 types of 2-DOF, 7-link, 2-basic loops generalized planar simple-jointed ($V = 0$) and multiple-jointed hinged mechanisms ($V = 1$, $V = 2$)

Note that all the synthesized 2-DOF, 7-link, 2-basic loops simple and multiple joint closed kinematic chains are guaranteed to present nonisomorphic different kinematic chains.

Furthermore the total result of structural synthesis (shown in Figs. 1 and 2) establishes that for 2-DOF, 7-link, 2-basic loops closed kinematic chains with $\alpha_{\text{min}} = \alpha_3$ there exist only 4 types of simple joint kinematic chains, 5 types of kinematic chains with 1 multiple joint (j_2), 2 types of kinematic chains with 2 multiple joints (j_2, j_2) and 1 type of kinematic chains with 1 multiple joint (j_3).

Then using the complete atlas of simple-jointed and multiple-jointed closed loop kinematic chains (presented in Figs. 1 and 2) and various permutations of fixed frame, we can construct the complete atlas database (shown below in Fig. 3) with all possible 2-DOF, 7-link, 2-basic loops mechanisms.

4 Creative Design of 2-DOF, 7-Link, 2-Basic Loops Hinged Robotic Mechanisms with Simple and Multiple Joints

If the complete atlas database of F-DOF, K-basic loop, V- total multiple joint factor kinematic chains with simple and multiple joints (Figs. 1 and 2) and corresponding mechanisms (Fig. 3) are established, the creative design of robotic mechanisms for a given task is the selection of appropriate topological structures of mechanisms from the complete atlas database according to the proposed design constraints. Design constraint usually includes the F-DOFs and the number of independent loop K (or the number of links) of mechanisms, value of V- total multiple joint factor, the type of simple ($V = 0$) and multiple ($V \geq 1$) joints, the type of the fixed frame, the type of input link and output link, etc. Different design constraints lead to different design results.

For example, if one seeks to find all feasible robotic mechanisms for creation of robot manipulators, the following design constraints can be given:

(1) The number of links for the required robotic mechanisms is 7 and the DOF of the mechanisms is 2.
(2) The fixed link (frame) is a multiple-pair binary or ternary link (as shown in Fig. 3).
(3) The topological structure of all candidate linkage for construction of 2-DOF, 7-link, 2-basic loops robotic mechanisms should be similar to the one in the atlases presented in Figs. 1, 2 and 3.
(4) The 2 input links with drive and output links (connected with end-effector as binary or ternary link) must perform the required complex motion in entire workspace.
(5) Robot manipulators can have single-arm or dual-arm work members.

In such a way, the examples of single-arm robot manipulators synthesized on the basis of the atlas of all feasible 2-DOF, 7-link, 2-basic loops simple-jointed and multiple-jointed kinematic chains (Figs. 1 and 2) and the atlas of mechanisms (Fig. 3) are constructed and presented in Fig. 4.

5 Mobility Analysis of Closed Simple-Jointed and Multiple-Jointed Kinematic Chains of Robotic Mechanisms

For mobility analysis of synthesized closed simple and multiple joint kinematic chains (shown in Figs. 1 and 2) and for its industrial applications in robot manipulators, we introduce a new mobility equation (degrees of freedom F) proposed by Pozhbelko [31]:

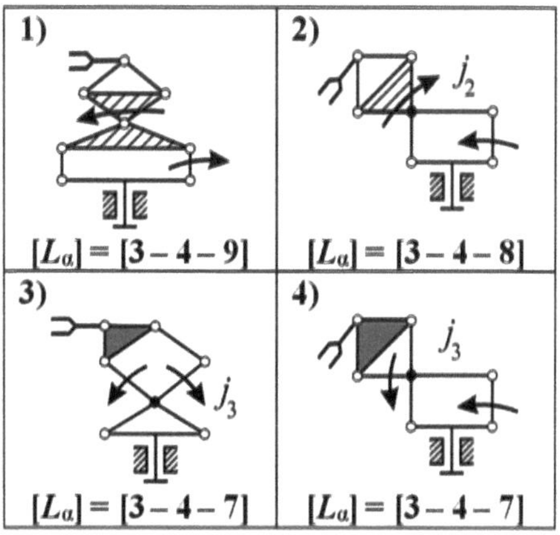

Fig. 4 Examples of 2-DOF, 7-link, 2-basic loops single-arm robot manipulators based on synthesized generalized kinematic chains and hinged mechanisms (presented in Figs. 1, 2 and 3)

$$F = \Sigma(3-i)n_i - V - 3 = (n_2 - 3) - \\ - [n_4 + 2n_5 + \cdots + (3 - i_{\max})n_{i_{\max}}] - V \qquad (4)$$

where total multiple joint factor (V) for closed kinematic chains can be expressed by following equation:

$$V = \Sigma(j-1)\nu_j \\ = \nu_2 + 2\nu_3 + \cdots + (j\max - 1)\nu_{j}\max \le 2(K-1) \qquad (5)$$

Example 1 For closed simple-jointed kinematic chain No. (1) (single-arm robot manipulator) shown in Fig. 4, we have original set of link assortment $[L.A.] = [n_2, n_3, n_4] = [5, 2, 0]$, multiple joint assortment $[M.J.A.] = [0]$ and loop assortment $[L_\alpha] = [\alpha_3 - \alpha_4 - \alpha_9] = [3 - 4 - 9]$, and the peripheral loop is $L_0 = 9$. For structural parameters $n_2 = 5$, $n_3 = 2$, $n_4 = 0$, $V = 0$ Eq. (4) yields $F = (n_2 - 3) - n_4 - V = (5 - 3) - 0 - 0 = +2$.

Example 2 For closed multiple-jointed kinematic chain No. (2) with 1 multiple joint (j_2) (single-arm robot manipulator) shown in Fig. 4, we have original set of link assortment $[L.A.] = [n_2, n_3, n_4] = [6, 1, 0]$, multiple joint assortment $[M.J.A.] = [\nu_2 \bullet \nu_3 \bullet \nu_4] = [1 \bullet 0 \bullet 0]$ and loop assortment $[L_\alpha] = [\alpha_3 - \alpha_4 - \alpha_8] = [3 - 4 - 8]$, and the peripheral loop is $L_0 = 8$. For structural parameters $n_2 = 6$, $n_3 = 1$, $n_4 = 0$, $V = \nu_2 + 2\nu_3 = 2$ Eq. (4) gives $F = (n_2 - 3) - n_4 - V = (6 - 3) - 0 - 1 = +2$.

Example 3 For closed multiple-jointed kinematic chain No. (3) with 1 multiple joint (j_3) (single-arm robot manipulator) shown in Fig. 4, we have original set of link assortment $[L.A.] = [n_2, n_3, n_4] = [7, 0, 0]$, multiple joint assortment $[M.J.A.] = [\nu_2 \bullet \nu_3 \bullet \nu_4] = [0 \bullet 1 \bullet 0]$ and loop assortment $[L_\alpha] = [\alpha_3 - \alpha_4 - \alpha_7] = [3 - 4 - 7]$, and the peripheral loop is $L_0 = 7$. For structural parameters $n_2 = 7$, $n_3 = 0$, $n_4 = 0$, $V = \nu_2 + 2\nu_3 = 2$ Eq. (4) gives $F = (n_2 - 3) - n_4 - V = (7 - 3) - 0 - 2 = +2$.

Example 4 For closed multiple-jointed kinematic chain No. (4) with 1 multiple joint (j_3) (single-arm robot manipulator) shown in Fig. 4, we have original set of link assortment $[L.A.] = [n_2, n_3, n_4] = [7, 0, 0]$, multiple joint assortment $[M.J.A.] = [\nu_2 \bullet \nu_3 \bullet \nu_4] = [0 \bullet 1 \bullet 0]$ and loop assortment $[L_\alpha] = [\alpha_3 - \alpha_4 - \alpha_7] = [3 - 4 - 7]$, and the peripheral loop is $L_0 = 7$. For structural parameters $n_2 = 7, n_3 = 0, n_4 = 0$, $V = \nu_2 + 2\nu_3 = 2$ Eq. (4) gives $F = (n_2 - 3) - n_4 - V = (7 - 3) - 0 - 2 = +2$.

6 Conclusion

In this paper, a new approach to structural synthesis and creative design of closed-loop mechanisms (on the basis of original addition synthesis method) is proposed. This proposed structural synthesis method permits to construct the complete atlas of all feasible 12 types of 2-DOF, 7-link, 2-basic loops kinematic chains (with the smallest closed loop $\alpha_{\min} = \alpha_3$) and corresponding mechanisms with simple and multiple joints.

Various examples of mechanical design of single-arm robot manipulators on the basis of synthesized simple and multiple joint 2-DOF, 7-link, 2-basic loop kinematic chains are established and illustrated to show the validity and effectiveness of the proposed structural synthesis method.

Furthermore, the mechanical creative design of various mechanisms based on proposed complete atlas makes it possible to access all feasible kinematic structures for creation of the most effective robot manipulators.

Acknowledgements The work was supported by Act 211 Government of the Russian Federation, contract № 02.A03.21.0011.

References

1. Erdman AG, Sandor G (1984) Mechanism design: analysis and synthesis. Prentice-Hall, Englewood Cliffs, New Jersey
2. Uicker JJ, Pennock GR, Shingley JE (2003) Theory of machines and mechanisms. Oxford University Press, New York
3. Hunt KH (1978) Kinematic geometry of mechanisms. Oxford University Press, New York
4. Crossley FRE (1964) A contribution to Gruebler's theory in the number synthesis of plane mechanisms. ASME J Eng Ind 86(1):1–8
5. Manolescu NI (1973) A method based on Baranov trusses and using graph theory to find the set of planar jointed kinematic chains and mechanisms. Mech Mach Theory 8(1):3–22
6. Butcher EA, Hartman C (2005) Efficient enumeration and hierarchical classification of planar simple-jointed kinematic chains: Application to 12-and 14-bar single degree-of-freedom chains. Mech Mach Theory 40(9):1030–1050
7. Dobrjanskyj L, Freudenstein F (1967) Some applications of graph theory to the structural analysis of mechanisms. ASME J Eng Ind 89(1):153–158
8. Tuttle ER, Peterson SW, Titus JE (1989) Enumeration of basic kinematic chains using the theory of finite group. J Mech Trans Autom 111(4): 498–503

9. Yan HS, Hwang YW (1990) Number synthesis of kinematic chains based on permutation groups. Math Comput Model 13(8):29–42
10. Hwang WM, Hwang YW (1992) Computer-aided structural synthesis of planar kinematic chains with simple joints. Mech Mach Theory 27(2):189–199
11. Yan HS, Chiu YT (2015) On the number synthesis of kinematic chains. Mech Mach Theory 89(7):128–144
12. Yan HS (1998) Creative design of mechanical devices. Springer, Singapore
13. Tsai LW (2001) Mechanism design: enumeration of kinematic structures according to function. CRC Press, Florida
14. Tsai LW (1999) Robot analysis: the mechanics of serial and parallel manipulators. Wiley, New York
15. Ceccarelli M (2004) Fundamentals of mechanics of robotic manipulation. Kluwer Academic Publishers, Dordrecht
16. Ghosal A (2006) Robotics: fundamental concepts and analysis. Oxford University Press, Oxford, UK
17. Angeles J 2007 Fundamentals of robotic mechanical systems. Theory, methods and algorithms. Springer, New York
18. Simoni R, Carboni AP, Martins D (2009) Enumeration of kinematic chains and mechanisms. Proc Inst Mech Eng Part C J Mech Eng Sci 223(4):1017–1024
19. Simoni R, Carboni AP, Simas H, Martins D (2011) Enumeration of kinematic chains and mechanisms review. In: 13th world congress in mechanism and machine science, Guanajuato, México, 19–25 June, 2011
20. Ding H, Hou F, Kecskemethy A, Huang Z (2011) Synthesis of a complete set of contracted graphs for planar non-fractionated simple-jointed kinematic chains with all possible DOFs. Mech Mach Theory 46(11):1588–1600
21. Ding HF, Huang P, Zi B, Kecskemethy A (2012) Automatic synthesis of kinematic structures of mechanisms and robots especially for those with complex structures. Appl Math Model 36(12):6122–6131
22. Ding HF, Hou FM, Kecskeméthy A, Huang Z (2012) Synthesis of the whole family of planar 1-DOF kinematic chains and creation of their atlas database. Mech Mach Theory 47(1):1–15
23. Ding HF, Zi B, Huang P, Kecskeméthy A (2013) The whole family of kinematic structures for planar 2-and 3-DOF fractionated kinematic chains. Mech Mach Theory 70(12):74–90
24. Yan HS, Hsu CH (1988) Contracted graphs of kinematic chains with multiple joints. Math Comput Model 10(9):681–695
25. Chu I, Cao W (1998) Systemics of Assur groups with multiple joints. Mech Mach Theory 33(8):1127–1133
26. Hsu CH (1992) Enumeration of basic kinematic chains with simple and multiple joints. J Franklin Inst 329(4):775–789
27. Ding H, Zhao J, Huang Z (2010) Unified structural synthesis of planar simple and multiple joint kinematic chains. Mech Mach Theory 45(4):555–568
28. Chu J, Zou Y (2014) An algorithm for structural synthesis of planar simple and multiple joint kinematic chains. Proc Inst Mech Eng Part C J Mech Eng Sci 228(12):2178–2192
29. Ermoshina E, Pozhbelko V (2017) Structural synthesis, mobility analysis and creation of complete atlas of multiloop planar multiple-jointed kinematic chains on base all possible sets of color multiple joints for industrial applications. In: Wenger P, Flores P (eds) New trends in mechanism and machine science–Theory and industrial applications. Mechanism and machine science, vol 43. Springer, pp 375–382
30. Pozhbelko V, Ermoshina E (2015) Number structural synthesis and enumeration process of all possible sets of multiple joints for 1-DOF up to 5-loop 12-link mechanisms on base of new mobility equation. Mech Mach Theory 90(8):108–127
31. Pozhbelko V (2016) A unified structure theory of multibody open, closed loop and mixed mechanical systems with simple and multiple joint kinematic chains. Mech Mach Theory 100(6):1–16

Structural Synthesis of 1-DOF up to 5-Loop Kinematic Chains with Multiple Joints and Its Analysis in Robotics

Ekaterina Kuts, Dmitry Kuts and Vladimir Pozhbelko

Abstract In this paper, the systematic method is developed to synthesize planar closed kinematic chains with all the possible kinds of multiple joints according to given K-independent loops and up to total multiple joint factor V_{max}. First, the structural synthesis method of multiple joint kinematic chains based on the combination of corresponding simple and multiple joints is presented. Then, the complete atlas database containing all valid multiple joints in closed fractionated kinematic chains with up to 5 independent loops and up to 12 links is established and illustrated. Further, corresponding all possible 14 types of multiple joint assortments up to $K = 5$ and up to total multiple joint factor $V_{max} = 8$ are obtained and classified in the tables for the first time. Next, the structural analysis of multiple-jointed fractionated kinematic chains with various total multiple joint factors is conducted, and main structural parameters are determinated. Finally, based on atlas database of synthesized kinematic chains with multiple joints, the examples of creation of robot manipulators are provided.

Keywords Structural synthesis · Multiple-jointed kinematic chains
Atlas database · Robot manipulators

1 Introduction

Structural synthesis is the first step in the creation of various mechanisms and machines and represents one of the most complex tasks in the mechanisms and machine theory [1, 2].

E. Kuts · D. Kuts · V. Pozhbelko (✉)
South Ural State University, (National Research University), Chelyabinsk, Russia
e-mail: pozhbelkovi@susu.ru

E. Kuts
e-mail: ekaterina.n.kuts@gmail.com

D. Kuts
e-mail: kutcda@susu.ru

R. Rizk and M. Awad (eds.), *Mechanism, Machine, Robotics and Mechatronics Sciences*, Mechanisms and Machine Science 58,
https://doi.org/10.1007/978-3-319-89911-4_15

Widely used synthesis methods are founded on: Assur groups, structural graphs, adjacency matrices, transformation of binary chains, finite and permutation groups, loop theory [3–14]. Mechanical systems may contain simple and multiple joints, but the use of multiple joints provides more simple construction and reduces the space requirement. At the same time structural synthesis of mechanical systems with multiple joint kinematic chains is difficult task [15–22].

Simoni et al. [11] obtained new results for the enumeration of kinematic chains that satisfy the general Gruebler mobility criterion and mechanisms without isomorphisms and degenerate chains using group theory and graph theory.

Ding et al. [14] presented a new classification of kinematic chains (non-fractionated chains and fractionated chains). Based on graph theory authors generated the whole family of the kinematic structures of 2- and 3-DOF fractionated planar kinematic chains.

Yan and Hsu [15] synthesized non-fractionated and fractionated kinematic chains with one and two multiple joints ($j = 2$ or $j = 3$) up to ten links and up to four independent loops based on the concept of contracted graph.

Ding et al. [16] proposed novel structural synthesis method of planar kinematic chains with multiple joints based on bicolor topological graph. Authors obtained 40 types of 8-link 1-DOF kinematic chains with one and two multiple joints and correspondingly 83 types of 9-link 2-DOF with one multiple joint.

Ding et al. [18] presented new bicolor topological graph and contracted graphs approach for structural synthesis of planar simple and multiple joint kinematic chains. The whole families of 1-DOF closed kinematic chains with one multiple joint and with up to 16 links were obtained.

Ding et al. introduced [19] an automatic method to synthesize planar non-fractionated kinematic chains with two multiple joints based on topological graphs.

Chu and Zou [20] on the basis of adding single-kinematic-chain method generated planar simple and multiple joints kinematic chains with specified DOF, the number of links up to total multiple joint factor $V_{\max} = 4$, i.e. the maximum number of multiple joints equals four.

The main aim of this paper is the structural synthesis of multiple-jointed kinematic chains and enumeration of all the possible multiple joint assortments ([M.J.A.]) for multiple-jointed kinematic chains up to maximum value of total multiple joint factor $V_{\max}$ with given K-independent loops. It can be used for the design of mechanisms with all possible multiple joint assortments.

We propose new notations "*Multiple Joint Assortment*" and "*Link Assortment*".

"*Multiple Joint Assortment*" ([M.J.A.]) is expressed as $[\text{M.J.A.}] = [\nu_2 \cdot \nu_3 \cdot \nu_4 \cdot \ldots \cdot \nu_j]$, where $\nu_2 \cdot \nu_3 \cdot \nu_4 \cdot \ldots \cdot \nu_j$ are the numbers of two-pin joints (j_2), three-pin joints (j_3), four-pin joints (j_4), …, j-pin joints ($j = j_{\max}$).

"*Link Assortment*" is expressed as $[L.A.] = [n_2, n_3, n_4, \ldots, n_i]$, where $n_2, n_3, n_4, \ldots, n_i$ are the number of binary, ternary, quaternary, …, i—nary links, etc., in order, having i joints incident to the link.

These new notations are necessary to form the multiple joint non-fractionated and fractionated kinematic chains.

Examples of the creation and complete structural analysis of various 1-DOF robot manipulators designed based on the atlas of synthesized multiple joint kinematic chains are provided for the first time.

1.1 Nomenclature

$\tilde{n}$—total number of links in the closed kinematic structure: $\tilde{n} = \sum_i n_i$;

n_i—the number of links with i joints in the closed kinematic chain, e.g. the number of single links (n_1), binary links (n_2), ternary links (n_3), quaternary links (n_4), pentagonal links (n_5), etc.;

ν_j—the number of multiple joints in the kinematic chain, every such joint must connects j + 1 links ($j \geq 1$), e.g. the number of two-pin joints is denoted by ν_2, correspondingly two-pin joint in figures is marked by j_2. In a similar way, the number of three-pin (ν_3), four-pin (ν_4), five-pin (ν_5) and six-pin (ν_6) joints are denoted.

V—total number of multiple joints in the kinematic chain $\left(V = v_2 + 2v_3 + \cdots + (j-1)\, v_j\right)$;

V—total multiple joint factor, for simple joint kinematic chains V is equal to 0, for multiple joint kinematic chains V is greater than 0;

$V_{\max}$—the maximum value of total number of multiple joints in the kinematic chain;

K—the number of independent closed loops in the kinematic chain;

F—the number of degrees of freedom of mechanical system related to the fixed link (DOF = F);

L_i—independent closed loop;

[*L.A.*]—link assortment in the kinematic chain;

[*M.J.A.*]—multiple joint assortment in the kinematic chain.

2 Structural Synthesis Method of Multiloop Multiple Joint Kinematic Chains

We develop a new approach to structural synthesis of multiple joint kinematic chains with all the possible kinds of multiple joints and with corresponding multiple joint assortments [M.J.A.] up to maximum total multiple joint factor ($V_{\max}$). Our method is based on the combination of corresponding simple and multiple joints. At first, we set the number of independent loops in the kinematic chain, it is an important feature of this method.

A new proposed method is unified for non-fractionated multiple joint kinematic chains (Steps 1–4) and fractionated multiple joint kinematic chains (Step 5).

This algorithm consists of the following steps (shown in Fig. 1):

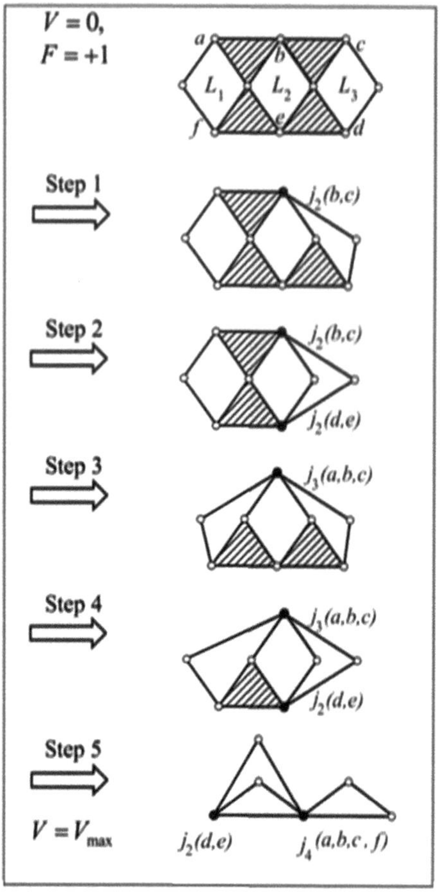

Fig. 1 The process to form multiple joint kinematic chains with given K-independent loops with $F = 1$

Step 1. Combining together simple joints (b) and (c) we obtain the first multiple joint j_2 (b, c).

Step 2. Combining together simple joints (d) and (e) we obtain the second multiple joint j_2 (d, e).

Step 3. Combining together simple joints (a), (b) and (c) we obtain the multiple joint j_3 (a, b, c).

Step 4. Combining together simple joints (a), (b), (c), (d) and (e) we obtain the multiple joints j_2 (d, e) and j_3 (a, b, c).

Step 5. Combining the various simple joints we obtain the multiple joints j_2 (d, e) and j_4 (a, b, c, f) up to $V = V_{\max}$.

3 Synthesis Results of Multiple-Jointed Kinematic Chains

According to the proposed algorithm of the structural synthesis for multiloop multiple-jointed kinematic chains the following results were received.

Figure 2 shows the synthesized 14 types 1-DOF fractionated multiple-jointed kinematic chains with up to 5 independent loops and up to 12 links with all the possible sets of various kinds of multiple joints up to maximum total multiple joint factor $V_{\max} = 8$.

Table 1 presents multiple joint assortment array for the structural analysis and synthesis of planar fractionated

1-DOF closed kinematic chains with up to 5 independent loops and up to 12 links with total multiple joint factor from $V = 1$ to $V = 8$.

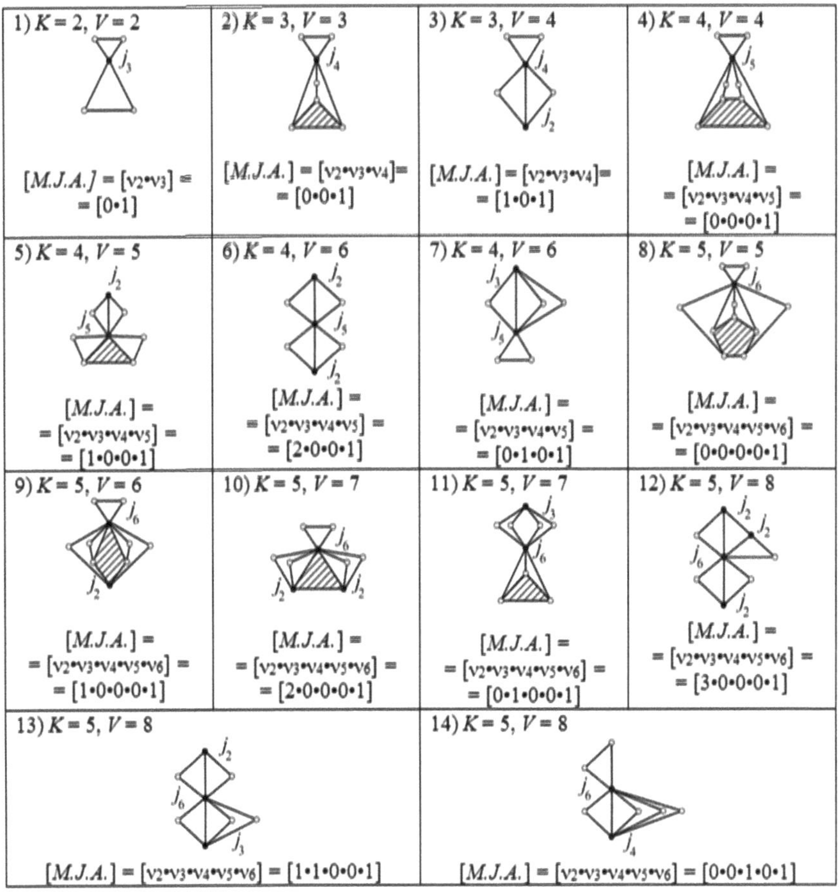

Fig. 2 The complete atlas of 1-DOF fractionated kinematic chains with up to 5 independent loops and up to 12 links with all possible multiple joint assortments

Table 1 Multiple joint assortment array of K-independent loops fractionated 1-DOF and $\tilde{n}$-link kinematic chains

$K=2$			$K=3$		
No.	V	[*M.J.A.*]	No.	V	[*M.J.A.*]
(1)	2	[0•1]	(2)	3	[0•0•1]
			(3)	4	[1•0•1]
$K=4$			$K=5$		
No.	V	[*M.J.A.*]	No.	V	[*M.J.A.*]
(4)	4	[0•0•0•1]	(8)	5	[0•0•0•0•1]
(5)	5	[1•0•0•1]	(9)	6	[1•0•0•0•1]
(6)	6	[2•0•0•1]	(10)	7	[2•0•0•0•1]
(7)	6	[0•1•0•1]	(11)	7	[0•1•0•0•1]
			(12)	8	[3•0•0•0•1]
			(13)	8	[1•1•0•0•1]
			(14)	8	[0•0•1•0•1]

Presented in Table 1 multiple joint assortment arrays can be used for synthesis and creation (based on any acceptable kinds of multiple joints) of the atlas database of all the possible multiloop multiple-jointed fractionated kinematic chains.

For example (for $K=4$), all 4 types of fractionated closed kinematic chains up to total multiple joint factor $V_{max}=6$ with any kinds of multiple joints can be synthesized from the corresponding "Multiple joint assortment array" (shown in Table 1).

It should be noted that the link assortments of the kinematic chain can be different for the same multiple joint assortments.

For example, the Fig. 3 shows 1-DOF $K=5$ multiple-jointed kinematic chains with following link assortments:

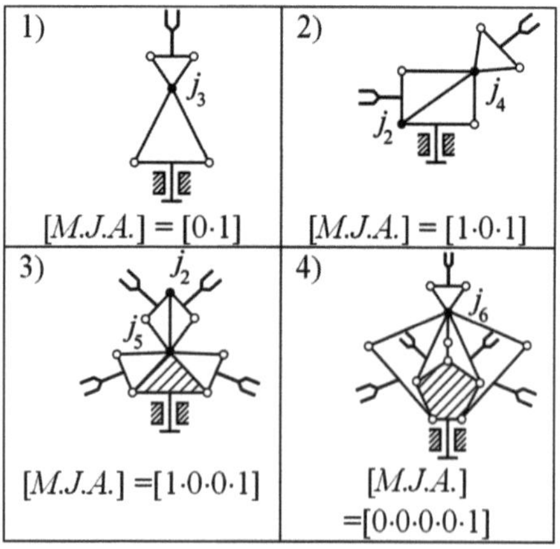

Fig. 3 The atlas of 1-DOF fractionated kinematic chains with 12 links $K=5$ fractionated kinematic chains with various link assortments

(1) $V = 5$, [$M.J.A.$] = [0•0•0•0•1], [$L.A.$] = [93000];
(2) $V = 5$, [$M.J.A.$] = [0•0•0•0•1], [$L.A.$] = [10.1100];
(3) $V = 5$, [$M.J.A.$] = [0•0•0•0•1], [$L.A.$] = [11.0010];
(4) $V = 6$, [$M.J.A.$] = [1•0•0•0•1], [$L.A.$] = [10.2000];
(5) $V = 6$, [$M.J.A.$] = [1•0•0•0•1], [$L.A.$] = [11.0100];
(6) $V = 7$, [$M.J.A.$] = [2•0•0•0•1], [$L.A.$] = [11.1000];

This algorithm (proposed in Fig. 1) has advantageous feature that distinguishes it from other algorithms: we can find different multiple joint assortments up to maximum total multiple joint factor ($V_{\max} = 8$). Thus, based on this algorithm we can obtain all the possible combinations of various kinds of multiple joints up to maximum total multiple joint factor $V_{\max}$.

The present paper provides a complete list of multiple joint assortment arrays for planar closed fractionated kinematic chains with up to 5 independent loops and up to 12 links with $V_{\max} = 8$ for the first time.

4 Directed Structural Analysis of Kinematic Chains with Maximum Total Multiple Joint Factor

We use the following four unified generalized equations proposed by Pozhbelko [22] for structural analysis of planar kinematic chains:

(1) Total multiple joint factor of a kinematic chain (V)

$$V = \sum (j-1)\nu_j \le n_1 + 2(K-1) \Rightarrow$$
$$V_{\max} = n_1 + 2(K-1) \quad (1)$$

(2) Degree of freedom (DOF) of a kinematic chain (F)

$$F = \left[\sum (3-i)n_i\right] - V - 3 \quad (2)$$

(3) Number of independent loops of a kinematic chain (K)

$$K = 1 + 0,5\left[V + \sum (i-2)n_i\right] \quad (3)$$

(4) Assembly criterion of a valid kinematic chain

$$\sum (i-2)n_i = 2(K-1) - V \quad (4)$$

On the base of the proposed Eqs. (1), (2), (3) and (4) we can carry out structural analysis of all closed multiloop fractionated kinematic chains with maximum total multiple joint factor $V = V_{max}(K)$ shown in Fig. 2.

Case 1. For 6-link closed fractionated kinematic chain shown in Fig. 2 No. (1), where $n_2 = 6$, $v_3 = 1$, the Eqs. (1), (2), (3) and (4) yield:

(1) $V = 2$; (2) $F = +1$; (3) $K = 2$; (4) $0 = 0$; 5) $V_{max} = 2$.

Case 2. For 8-link closed fractionated kinematic chain shown in Fig. 2 No. (3), where $n_2 = 8$, $v_2 = 1$, $v_4 = 1$, the Eqs. (1), (2), (3) and (4) yield:

(1) $V = 4$; (2) $F = +1$; (3) $K = 3$; (4) $0 = 0$; (5) $V_{max} = 4$.

Case 3. For 10-link closed fractionated kinematic chain shown in Fig. 2 No. (6), where $n_2 = 10$, $v_2 = 2$, $v_5 = 1$, the Eqs. (1), (2), (3) and (4) yield:

(1) $V = 6$; (2) $F = +1$; (3) $K = 4$; (4) $0 = 0$; (5) $V_{max} = 6$.

Case 4. For 12-link closed fractionated kinematic chain shown in Fig. 2 No. (14), where $n_2 = 12$, $v_4 = 1$, $v_6 = 1$, the Eqs. (1), (2), (3) and (4) yield:

(1) $V = 8$; (2) $F = +1$; (3) $K = 5$; (4) $0 = 0$; (5) $V_{max} = 8$.

5 Industrial Application of Synthesized Multiloop Kinematic Chains in Robotics

Synthesized fractionated multiple-jointed kinematic chains, shown in Fig. 2, can be used in various fields of mechanical engineering, for example, in various robot manipulators, shown below in Fig. 4.

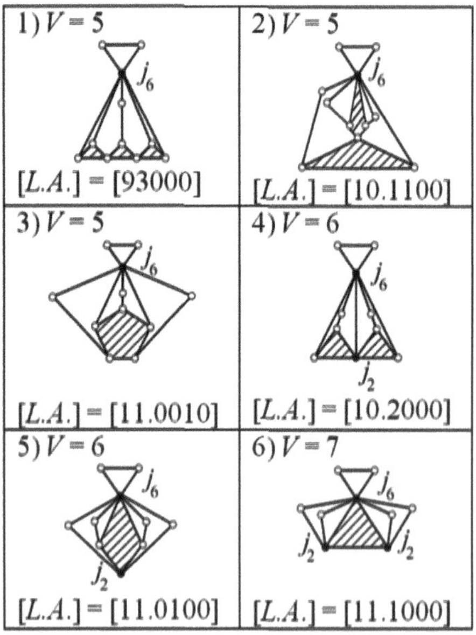

Fig. 4 Industrial application of fractionated multiple-jointed kinematic chains in robotics

Based on the proposed unified Eqs. (1), (2), (3) and (4) we can carry out structural analysis of robot manipulators shown in Fig. 4.

Case 1. For robot manipulator (6-link closed kinematic chain from Fig. 2, No. 1) shown in Fig. 4 No. (1), where $n_2 = 6$, $\nu_2 = 0$, $\nu_3 = 1$, the Eqs. (1), (2), (3) and (4) yield:

(1) $V = 2$; (2) $F = +1$; (3) $K = 2$; (4) $0 = 0$; (5) $V_{\max} = 2$.

Case 2. For robot manipulator (8-link closed kinematic chain from Fig. 2, No. 3) shown in Fig. 4 No. (2), where $n_2 = 8$, $\nu_2 = 1$, $\nu_4 = 1$, the Eqs. (1), (2), (3) and (4) yield:

(1) $V = 4$; (2) $F = +1$; (3) $K = 3$; (4) $0 = 0$; (5) $V_{\max} = 4$.

Case 3. For robot manipulator (10-link closed kinematic chain from Fig. 2, No. 5) shown in Fig. 4 No. (3), where $n_2 = 9$, $n_3 = 1$, $\nu_2 = 1$, $\nu_5 = 1$, the Eqs. (1), (2), (3) and (4) yield:

(1) $V = 5$; (2) $F = +1$; (3) $K = 4$; (4) $0 = 0$; (5) $V_{\max} = 6$.

Case 4. For robot manipulator (12-link closed kinematic chain from Fig. 2, No. 8) shown in Fig. 4 No. (4), where $n_2 = 11$, $n_5 = 1$,$\nu_2 = 6$, $\nu_3 = 0$, the Eqs. (1), (2), (3) and (4) yield:

(1) $V = 5$; (2) $F = +1$; (3) $K = 5$; (4) $0 = 0$; (5) $V_{\max} = 8$.

6 Multiple Joint Closed Kinematic Chains Theorem

Using the above results of the structural synthesis and analysis (illustrated in Fig. 2 for $F = 1$ and the number of independent loops up to five), we establish the following theorem.

Theorem *All the possible F-DOF and K-independent loops multiple-jointed closed kinematic chains with given $j_{max} > K$ can exist only as fractionated kinematic chains in synthesis process of hinged mechanical systems with structural parameters:*

$$jmax > K \tag{5}$$

as a criterion of existence of fractionated multiple joint kinematic chains.

Proof Various kinematic chains can be transformed to generalized mechanical systems which consist of only binary links [22]. Using the graph theory, we prove this theorem for systems consisted only of binary links (shown in Fig. 2, No. 14).

Let us consider the graph corresponding to the closed kinematic chain. The vertices of the graph are in one-to-one correspondence with the joints of the kinematic chain, the edges are in one-to-one correspondence with the links of the kinematic chain, respectively. This graph has N vertices, M edges and one connected component. The vertices of degree 0 and 1 are absent in such graph.

In our case, according to [23], we have $K = M - N + 1$, where K is the number of independent loops. By condition, we have that there is the vertex ε whose degree is equal to $\rho = j_{max} + 1$. Therefore, the condition of our theorem takes the following form $\rho > K + 1 = M - N + 2$ or $\rho \geq M - N + 3$.

Let us prove by contradiction that if $\rho \geq M - N + 3$, then our kinematic chain must be fractionated. Let us assume that our kinematic chain is not fractionated. If we remove the vertex ε with its incident edges, we get a new graph, with $M_{new} = M - \rho$ and $N_{new} = N - 1$. This new graph has the number of independent loops, which is equal to $K_{new} = M_{new} - N_{new} + 1 = M - N + 2 - \rho$. Combining the last equation and the inequality $\rho \geq M - N + 3$, we get $K_{new} < 0$. This leads to a contradiction. The theorem is proved.

7 Conclusion

In this paper, we have proposed structural synthesis method based on the combination of corresponding simple and multiple joints.

The complete atlas database of closed kinematic chains with up to 5-independent loops and up to 12-link, up to pentagonal links and up to total multiple joint factor $V_{max} = 8$ is established and demonstrated. As a result, the complete list of all possible multiple joint assortments for fractionated multiple joint kinematic chains is obtained.

Examples of structural analysis of various $\tilde{n}$-link, 1-DOF kinematic chains with various total multiple joint factors for industrial applications in robotics are presented. Efficiency and effectiveness of proposed unified equations for topological structure and mobility analysis of various robot manipulators are shown.

The methodology of structural synthesis and analysis developed in this paper can be helpful for designers to create a complete catalog of multiple-jointed fractionated multiloop closed kinematic chains in the process of the creative design of mechanisms.

Acknowledgements The work was supported by Act 211 Government of the Russian Federation, contract № 02.A03.21.0011. The authors thank Associate Professor Aleksander Evnin for fruitful discussion on the graph theory.

References

1. Norton RL (2003) Design of machinery: an introduction to the synthesis and analysis of mechanisms and machines. McGraw-Hill, Boston
2. Uicker JJ, Pennock GR, Shingley JE (2003) Theory of machines and mechanisms. Oxford University Press, New York
3. Crossley FRE (1964) A contribution to Gruebler's theory in the number synthesis of plane mechanisms. ASME J Eng Industry 86(1):1–8
4. Dobrjanskyj L, Freudenstein F (1967) Some applications of graph theory to the structural analysis of mechanisms. ASME J Eng Industry 89(1):153–158
5. Yan HS, Hwang YW (1990) Number synthesis of kinematic chains based on permutation groups. Math Comput Modelling 13(8):29–42

6. Mruthyunjaya TS (1979) Structural synthesis by transformation of binary chains. Mech Mach Theory 14:221–231
7. Yan HS, Chiu YT (2015) On the number synthesis of kinematic chains. Mech Mach Theory 89(7):128–144
8. Yan HS (1998) Creative design of mechanical devices. Springer Verlag, Singapore
9. Tsai LW (1999) Robot analysis: the mechanics of serial and parallel manipulators. Wiley-Interscience, New York
10. Ceccarelli M (2004) Fundamentals of mechanics of robotic manipulation. Kluwer Academic Publishers, Dordrecht
11. Simoni R, Carboni AP, Martins D (2009) Enumeration of kinematic chains and mechanisms. Proc Inst Mech Eng Part C J Mech Eng Sci 223(4):1017–1024
12. Butcher EA, Hartman C (2005) Efficient enumeration and hierarchical classification of planar simple-jointed kinematic chains: application to 12- and 14-bar single degree-of freedom chains. Mech Mach Theory 40(9):1030–1050
13. Ding HF, Huang P, Zi B, Kecskemethy A (2012) Automatic synthesis of kinematic structures of mechanisms and robots especially for those with complex structures. Appl Math Modelling 36(12):6122–6131
14. Ding HF, Zi B, Huang P, Kecskeméthy A (2013) The whole family of kinematic structures for planar 2-and 3-DOF fractionated kinematic chains. Mech Mach Theory 70(12):74–90
15. Yan HS, Hsu CH (1988) Contracted graphs of kinematic chains with multiple joints. Math Comput Modelling 10(9):681–695
16. Ding H, Zhao J, Huang Z (2010) Unified structural synthesis of planar simple and multiple joint kinematic chains. Mech Mach Theory 45(4):555–568
17. Hsu CH (1992) Enumeration of basic kinematic chains with simple and multiple joints. J Frankl Inst 329(4):775–789
18. Ding H, Yang W, Huang P, Kecskeméthy A (2012) Automatic structural synthesis of planar multiple joint kinematic chains. ASME J Mech Des, 135(9): 031006-1–031006-10
19. Ding H, Yang W, Zi B, Kecskeméthy A (2016) The family of planar kinematic chains with two multiple joints. Mech Mach Theory 99(5):103–116
20. Chu J, Zou Y (2014) An algorithm for structural synthesis of planar simple and multiple joint kinematic chains. Proc Inst Mech Eng Part C J Mech Eng Sci 228(12): 2178–2192
21. Pozhbelko V, Ermoshina E (2015) Number structural synthesis and enumeration process of all possible sets of multiple joints for 1-DOF up to 5-loop 12-link mechanisms on base of new mobility equation. Mech Mach Theory 90(8):108–127
22. Pozhbelko V (2016) A unified structure theory of multibody open, closed loop and mixed mechanical systems with simple and multiple joint kinematic chains. Mech Mach Theory 100(6):1–16
23. Berge C (2001) The theory of graphs. Dover Publications, Mineola, New York

New Cable Driven Wrist

E. Backri Kasbah, M. Rizk, C. Kassis and R. Rizk

Abstract Inertia is one of the most annoying problems in robotics. Cable driven mechanisms are characterized by their low inertia. However, when an epicyclic train is used, kinematics becomes very complicated. This paper shows a wrist, driven by a pulley tendon system. A generalized form for the Willis formula is developed. The pulley arrangement is done, such as, there is a direct driving between the fixed actuator on the frame and the link connected to the last pulley.

Keywords Wrist · Cable driven mechanism · Differential mechanism

1 Introduction

Most robotics applications require wrists. Cutting, polishing welding etc. require a simultaneous control of position and orientation, i.e. a wrist. It is the combination of several rotational motions. Complex non-inertial effects, centrifugal and Coriolis, appear. Inertia induces vibrations and hard control. Light weight is a must. It is not possible to eliminate the link weight. The actuator weight can be removed by the power transmission [1]. This can be achieved using gears, linkages or cables. Several Orthosis [2], grippers [3, 4], arms [1] and wrists [5] with power transmission are suggested in literature. Those devices can be classified based on their transmission mechanisms:

E. B. Kasbah · M. Rizk · C. Kassis (✉)
Faculty of Engineering, CRSI, Lebanese University, Beirut, Lebanon
e-mail: carine.kassis@ul.edu.lb

E. B. Kasbah
e-mail: elise.kasbah@smartageeng.com

M. Rizk
e-mail: maher-rizk@hotmail.com

R. Rizk
Faculty of Engineering, CRSI, Lebanese University, Roumieh, Lebanon
e-mail: rrizk2@ul.edu.lb

R. Rizk and M. Awad (eds.), *Mechanism, Machine, Robotics and Mechatronics Sciences*, Mechanisms and Machine Science 58,
https://doi.org/10.1007/978-3-319-89911-4_16

a. Cable driven systems. This category can be divided into two subcategories, Cable-driven parallel systems [2–8] and cable-pulley systems [1, 9–12]. Systems of this category are characterized by low inertia and fast response. However they need high manufacturing accuracy, pretension in the cables, and a complex motion since they induce high coupling.
b. Gear driven systems [13, 14]. System of this category can be used for heavy tasks. They can support high load. However, they do not decrease significantly the inertia, they present high backlash, and induce high coupling.
c. Linkage systems [15–23]. They support heavy loads, have lower backlash than gear systems. However they are heavier and still induce coupling.

The main problem of previous systems is the complicated kinematics. The classical formulas of Denavit-Hartenberg cannot be used since there is no direct driving between the actuator and the link. The rotation of the driven link becomes relative to the global frame instead the folding angle. In this paper, we are looking for a cable-pulley system that provides direct driving between the actuator, installed on the frame, and the folding angle of the arm.

On the other hand, most wrists presented in literature have a wrist point different than the operational point. If the robot works with a fluid, the acceleration of the operational point leads to the cone and taper effect [24]. This is due to the deflection of the fluid stream. For abrasive waterjet cutting, on straight cuts, the nozzle can be moved swiftly since the deflection does not affect the accuracy [25]. However, deflection on corners can cause cutting errors. This problem is mainly due to the water stream inertia. This can be solved by using a wrist point coincident with the operational point. The wrist point should be external. Axes of the revolute joints have to intersect at the same point with the water stream.

This paper presents a wrist used for a marble water jet cutting machine, driven by a cable pulley system. In Sect. 2 the kinematics structure of the wrist is explained. In Sect. 3, mechanisms used to ensure a direct driving between the actuator and the folding angle are developed and presented. In Sect. 4 a prototype is shown. We finish our paper by conclusions and further challenges.

2 Kinematic Structure

A water jet is a five-axis machine. The base is Cartesian with three linear actuators. The tip has to be coincident with the wrist point. For this reason, axes 4 and 5 must have an external intersection point. We need the kinematic skeleton shown in Fig. 1. This allows changing the orientation of the cutting stream without any control on the position. Moreover, this induces the minimization of the cone and taper effect. The cutting point velocity is zero at corners.

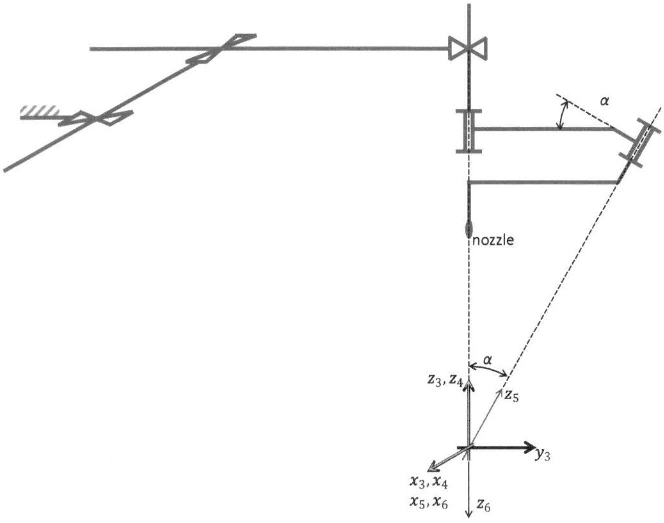

Fig. 1 Kinematics scheme of the system

2.1 *Wrist Kinematics Model*

The torsion angle of link 4 is $\alpha = 30^{\circ}$. This value allows to engrave until an inclination of $\pm 60^{\circ}$, which is the maximum needed in marble manufacturing. The nozzle axis passes through the intersection point of joints 4 and 5. This allows engraving a perfect axisymmetric shape, simply by looking the first three joints. In the rest of the paper, we are interested in the wrist only.

Since light weight is needed, we have to use cable power transmission. Actuators 4 and 5 are hold by link 3. The power transmission is needed between motor 5 and the corresponding link. In the following a planetary system, with an angle α between the axes of the sun and the planet, is used. The relationship connecting the angular motion of those two gears has to be developed. The Willis formula [4] is used when the axes are parallel. In the following we are going to derive this relationship with an angle between the axes, for a gear system as well as for a pulley tendons system. Later, this relationship is used to develop the train that generates a direct driving.

2.2 *Kinematics Relationship for Planetary Gear Train*

Figure 2 shows a planetary train of two gears carried by an arm a with an angle α between the gear axes. Gear 2 is rotating with an angular velocity ω_2, Gear 3 with an angular velocity ω_3, and the arm with an angular velocity ω_a. Three angular velocities

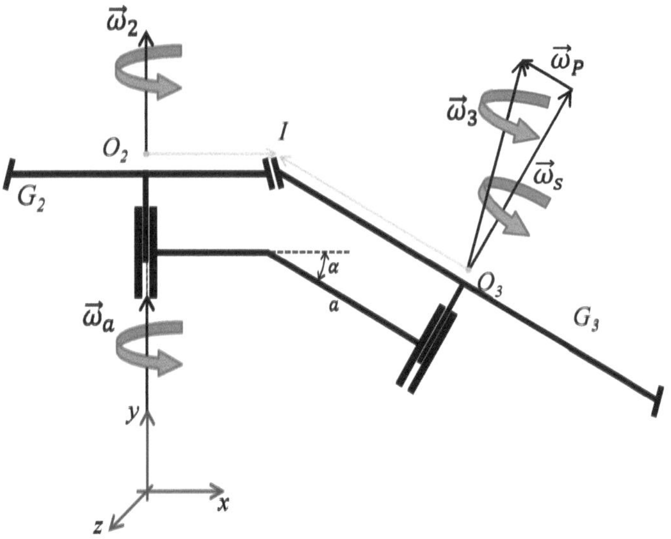

Fig. 2 Planetary gear train with inclined axes

are described in the global reference frame. The angular velocity $\vec{\omega}_3$ can be seen as the result of its spinning $\vec{\omega}_S$ and its precession $\vec{\omega}_P$:

$$\vec{\omega}_3 = \vec{\omega}_S + \vec{\omega}_P \tag{1}$$

The velocity of the contact point I is the same on both gears. Seen from gear 2, the velocity of point I is:

$$\vec{v}_I = \vec{\omega}_2 \times \overrightarrow{O_2 I} = \omega_2 \vec{j} \times r_2 \vec{i} = -\omega_2 r_2 \vec{k} \tag{2}$$

Seen from gear 3, the velocity $\vec{v}_I$ is:

$$\begin{aligned} \vec{v}_I &= \vec{\omega}_a \times \overrightarrow{O_2 O_3} + \vec{\omega}_3 \times \overrightarrow{O_3 I} \\ &= \omega_a \vec{j} \times \left((r_2 + r_3 \cos\alpha)\vec{i} - r_3 \sin\alpha \vec{j} \right) \\ &+ \left[\omega_P \left(-\cos\alpha \vec{i} + \sin\alpha \vec{j} \right) + \omega_S \left(\sin\alpha \vec{i} + \cos\alpha \vec{j} \right) \right] \\ &\times r_3 \left(-\cos\alpha \vec{i} + \sin\alpha \vec{j} \right) \\ &= \left[-\omega_a (r_2 + r_3 \cos\alpha) + \omega_S r_3 \right] \vec{k} \end{aligned} \tag{3}$$

By equating Eqs. (2) and (3):

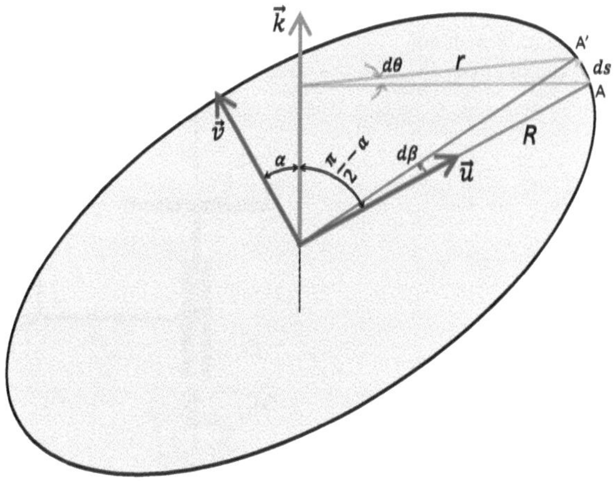

Fig. 3 Rotation of a disc around an deviated axis

$$\frac{\omega_2 - \omega_a}{\omega_S - \omega_a \cos\alpha} = -\frac{r_3}{r_2} \tag{4}$$

Equation (4) can be considered as a general form of the Willis formula. We will prove that this form is also valid for a pulley-tendon system. A disc (Fig. 3) with an axis $\vec{v}$ is turned around an axis $\vec{k}$, such as the angle between axes $\vec{v}$ and $\vec{k}$ is α, with an angular velocity $\dot{\theta}$. It is known [26]:

$$d\theta\vec{k} = d\theta \sin\alpha\vec{u} + d\theta \cos\alpha\vec{v} = d\theta \sin\alpha\vec{u} + d\beta\vec{v} \tag{5}$$

where, $d\beta = d\theta.\cos\alpha$ is the rotation of the disc around its own axis $\vec{v}$.

For a rotation $d\theta$ around $\vec{k}$, the displacement of a point A at the periphery of the disc is:

$$\overrightarrow{AA'} = d\theta\vec{k} \times R\vec{u} \tag{6}$$

Then:

$$AA' = ds = r.d\theta = R\cos\alpha d\theta = Rd\beta \tag{7}$$

The distance *ds* is the length of the tendon enrolled on the pulley when the pulley is turned by an angle $d\theta$ around the axis $\vec{k}$.

In the system shown in Fig. 4, if the arm is rotating with an angular velocity ω_a, pulley P_2 with ω_2, the question is how much is ω_3?

Let the angular velocity of P_3 resulted from the rotation of the arm be ω_{3a} and ω_{32} the one resulted from P_2. ω_3 is the summation of the results. First, if the arm is blocked and P_2 is actuated, the same length of the cable is enrolled on P_3 and P_2. Then:

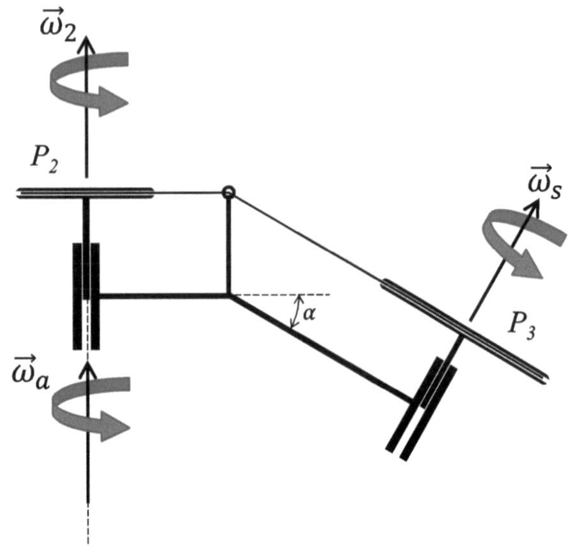

Fig. 4 Planetary pulley tendon train with non co-axial pulleys

$$\omega_2 r_2 = \omega_{32}\, r_3 \tag{8}$$

If the arm is actuated and P_2 is blocked, the phenomenon is equivalent to rotating whole the device with an angular velocity ω_a, and P_2 is rotating with an angular velocity $-\omega_a$. The resulted angular velocity ω_{3a} is:

$$\omega_{3a} = \omega_a \cos\alpha - \frac{r_2}{r_3}\omega_a \tag{9}$$

Finally the angular velocity is:

$$\omega_3 = \omega_a \cos\alpha + \frac{r_2}{r_3}\omega_2 - \frac{r_2}{r_3}\omega_a = \left(\cos\alpha - \frac{r_2}{r_3}\right)\omega_a + \frac{r_2}{r_3}\omega_2 \tag{10}$$

And the formula is similar to Eq. (4):

$$\frac{\omega_2 - \omega_a}{\omega_3 - \omega_a \cos\alpha} = \frac{r_3}{r_2} \tag{11}$$

ω_3 is the angular velocity of the pulley 3 around its own axis. We could develop a generalized form for the Willis formula. We have to use this formula in order to generate a direct driving, between the actuator and the inclined link.

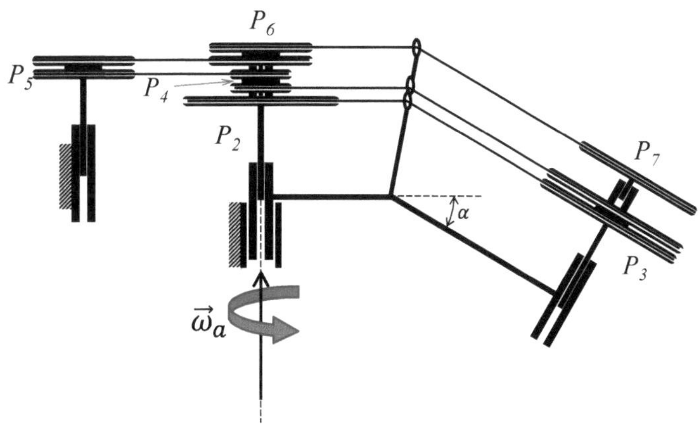

Fig. 5 Transmission mechanism skeleton

2.2.1 Power Transmission Mechanism

The goal is to get a behavior for the second link similar to that in a robotic arm. If ω_2 is the angular velocity of the actuator, ω_a that of the arm and ω_L the spinning velocity of the second link, we need:

$$\omega_2 = \omega_L - \omega_a \cos \alpha \tag{12}$$

To get the mechanism generating Eq. (12) mainly two particular ratios are needed. Ratio R_i that ensures a driving independent of the link motion, and a ratio R_d that ensures a direct driving or at least respects the ratio between ω_2 and ω_a of Eq. (12). From Eq. (10), to get a driving independent of ω_a, the coefficient of ω_a should be zero thus:

$$R_i = \frac{r_2}{r_3} = \cos \alpha \tag{13}$$

To get a driving between pulley 2 and the arm similar to Eq. (12), using Eq. (10):

$$\frac{\cos \alpha - \frac{r_2}{r_3}}{\frac{r_2}{r_3}} = \cos \alpha \Rightarrow R_d = \frac{r_2}{r_3} = \frac{\cos \alpha}{1 + \cos \alpha} \tag{14}$$

In the system shown in Fig. 5, P_2 and P_3 have the ratio $\frac{r_2}{r_3} = \frac{\cos \alpha}{1+\cos \alpha}$. Substituting in Eq. (10):

$$\frac{1 + \cos \alpha}{\cos \alpha} \omega_3 = \omega_2 + \omega_a \cos \alpha \tag{15}$$

ω_3 has to be reduced by a coefficient independent of ω_a. P_3 is a double neck pulley. Pulley P_4 is driven by P_3 and its angular velocity has to be independent of the arm velocity, then $\frac{r_4}{r_3} = \cos\alpha$. Substituting in Eq. (10):

$$\omega_3 = \cos\alpha.\omega_4 \tag{16}$$

Equations (15) and (16) give:

$$(1 + \cos\alpha)\,\omega_4 = \omega_2 + \omega_a \cos\alpha \tag{17}$$

The goal is $\omega_L = \omega_7 = \omega_2 + \omega_a \cos\alpha$, (Eq. 12) then

$$\omega_7 = (1 + \cos\alpha)\,\omega_4 \tag{18}$$

Pulley 7 holds the second link. To reach the goal, an external reduction, and a driving independent of the arm velocity are needed to get the desired velocity. Two pulleys P_6 and P_7 such as $\frac{r_6}{r_7} = \cos\alpha$ give:

$$\omega_7 = \omega_6 \cos\alpha \tag{19}$$

Equations (18) and (19) give:

$$\omega_6 = \frac{1 + \cos\alpha}{\cos\alpha}\omega_4 \tag{20}$$

But

$$\frac{\omega_6}{\omega_4} = \frac{\omega_6}{\omega_5}\frac{\omega_5}{\omega_4} = \frac{r_4}{r_5}\frac{r_5}{r_6} = \frac{1 + \cos\alpha}{\cos\alpha} \tag{21}$$

P_5 is an idler, any radius for P_5 can be used. The problem solution is a set of pulleys with radii verifying previous equations. There are 4 constraints defined by Eqs. (15), (16), (19) and (21), with 5 parameters. Since the equations are set up in this section, we have to use them in order to design the prototype.

3 Mechanical Design and Prototype

A prototype is necessary to prove the correctness of the design, especially the transmission. The pitch/yaw system prototype (Fig. 6) is mounted on a chassis that will restrict the rotation of the first link 180°, while allowing a complete rotation for the second link. The chassis will serve only as a support for the system. In reality, the latter will be mounted on the z axis and the two links will be able to rotate 360°.

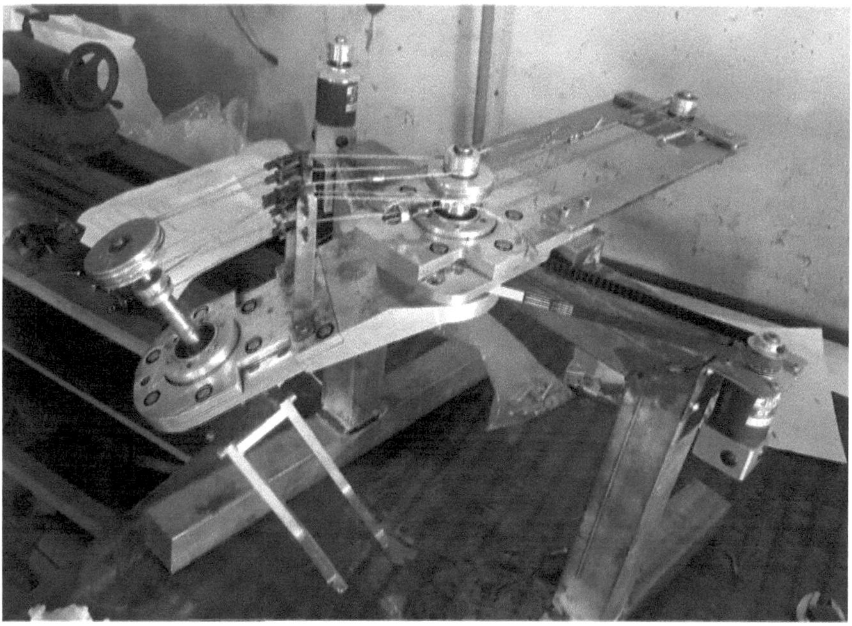

Fig. 6 Prototype

Details of manufactured parts (pulleys, shafts), and assembly are explained in the following.

3.1 Pulleys

Links dimensions are imposed by the industrial application. The angle α is equal to 30°. The pulley radii have to verify Eqs. (15), (16), (19) and (21). Moreover Pulleys 3, 4, 5 and 6 are double neck. The same radius for both necks in each pulley is used. Since there are 4 constraints and 5 parameters, the diameter of P_2 is chosen 25 mm. For P_5, a diameter of 25 mm is also used. Pulley diameters are shown in Table 1

Pulleys are made by Alumec 89 known for its low density. The cable is turned to form a closed loop similar to a belt. Both extremities are attached via a simple link. Increasing the number of turns of the cable around the pulley increases the precision, the tension in the cable and the friction. Compromising between all these criteria, it was decided to turn the cable 1.5 turns around each pulley.

Table 1 Power transmission system pulleys

Pulley	2,5, 7	3	4	6
Diameter (mm)	25	53.86	46.25	21.65

Fig. 7 Pulleys used in the transmission mechanism

All the pulleys are manufactured. It was better to thread the whole thickness of the pulleys. This effect allows to use the desired number of turns as well as to use it as a single or double pulley (Fig. 7). In addition, the main reason of threading the whole thickness is to prevent the different turns of the same cable to cross each other.

All the pulleys have an internal diameter of 13 mm, in which a bronze bushing is inserted by pressing. The bronze bushing is used for mounting the pulleys on the shafts while allowing them to rotate freely. A lubricant reduces more the friction. A spacer of 1 mm is added between the pulleys mounted on the second shaft, in order to prevent the friction between each other (Fig. 8).

3.2 *Shafts*

Shaft 1 is used for holding pulley 5. It is not joined to any link rotation. The shaft has a base of 36 mm diameter and 8 mm height. The pulley has an internal diameter of 13 mm, but taking into consideration the bronze ring, it was reduced to 11 mm. The length of this part is equal to the plate's thickness of 10 mm in addition to pulley's thickness of 15 mm and the circlip thickness of 3 mm used to prevent the axial motion of the axis. Hence its total length is:

$$l_1 = 8 + 10 + 15 + 3 = 36\,\text{mm} \tag{22}$$

Fig. 8 Shaft 1

The middle shaft is used to rotate the first link. It is connected to a timing belt pulley. In addition, it is used to hold the middle set of pulleys in the power transmission mechanism. (Figure 9)

The timing belt pulley is installed between the fixed plate and the first link. It has an internal diameter of 24 mm and a thickness of 15 mm. Hence, the following stages of the shaft have these dimensions.

The fixed plate has a thickness of 26 mm, in which two bearings of thickness 11 mm each are fitted. They have an internal diameter of 15 mm. To hold these two bearings in place, a shoulder is made among the lowest one and the timing belt pulley, and two retaining rings are used at each side of the second bearing.

Additionally, three pulleys of thickness 13 mm each are mounted on the axis and are free to rotate around it. Spacers are placed between the pulleys to prevent them from scratching against each other and increasing the friction. Finally a retaining ring is used to avoid the slipping of the pulleys axially.

The inclined shaft is used to rotate the second link. It is fixed to the link by screws and to pulley 7 by a key (Fig. 10). The base has a diameter of 60 mm and a height of 10 mm. The diameter is reduced to 35 mm in order to fill the gap already existing of thickness 10 mm.

Similarly to the middle axis, two bearings are installed in the same way. Thus, the diameter becomes 15 mm. In order to mount the pulleys, the diameter is reduced

Fig. 9 Middle shaft

further to 11 mm. Finally, the length of the shaft and the position of the pulleys were calculated in a way that keeps a 30° angle between the middle axis and the inclined axis.

3.3 *Assembly*

Finally, after joining all the previous parts, we get the assembly shown in Fig. 6

In order to fold cables, metallic housing for each cable is used. Finally a special tensioner is used to forbid the sliding between cables and pulleys (Fig. 6).

Fig. 10 Inclined shaft

4 Conclusions and Further Works

In this paper, we developed a relationship for the angular motion of a planetary system with an angle transmission. It can be used for a gear or pulley-cable system. This formula can be considered as a generalized form of the Willis formula. Using it, we can develop a power transmission system with an isotropy [27] between the folding angle and the intermediate link motion. The mechanism contains seven pulleys but it is very light, and it reduces considerably the inertia of the mechanism. The isotropy between the actuator and the folding angle allows using the formulas of D-H parameters and the classical robotics calculations without any limitation. To validate our calculation we manufactured a prototype for the wrist with an external tip. The industrial application in this paper was the abrasive water jet, however it can be used for any other application. We presented the manufacturing and the assembly of all the parts used in the prototype. It can be mounted on the Cartesian machine without any limitation.

In this paper, we considered our system as perfectly manufactured and assembled, but in a next step, we have to identify the assembly and manufacturing errors and compensate for them. We used only two links. The problem becomes more complicated if more links are used. It will be interesting to develop a complete arm with an isotropic transmission system. Mostly the cosine of the torsion angle is going to be zero if this angle is right. In the case of a right angle, the problem becomes more challenging. Moreover, we ignored the friction effect and the elasticity of the tendons. The backlash and the deflection of the shafts can also create considerable problems. All these effects are going to be the subjects of our next works.

References

1. Jr. Salisbury JK et al. (1993) Compact cable transmission with cable differential, US Patent, US5207114, May, 4, 1993
2. Streit DA, Gilmore BJ (1989) Perfect spring equilibrators for rotatable bodies, ASME Trans J Mech Transmissions Automation Des (111) 451–458
3. Jacobsen SC, Iversen EK, Knutti DF, Johnson RT, Biggers KB (1986) Design of the UTAH/MIT dextrous hand. In Proceedings of ICRA86, San Francisco, CA, USA, pp. 1520–1532
4. Laliberté T, Birglen L, Gosselin C (2002) Under-actuation in robotic grasp hands. Japanese J Mach Intell Robot Control 4(3):77–87 Special Issue on Under-actuated Robots
5. Bartholet SJ (1992) Reconfigurable End Effector, U.S. Patent 5 108 140
6. Ebert-Uphoff I, Johnson K (2002) Practical considerations for the static balancing of mechanisms of parallel architecture, J Multi-body Dynamics 216(K): 73–85
7. Nathan U, Kumar V (1991) Passive mechanical gravity compensation for robot manipulators. In: Proceedings of ICRA'91, California, USA
8. Dandache G, Rizk R, Krut S, Dombre E (2011) A pseudo-isotropic three phalanx under-actuated finger. In: 13th World congress in mechanism and machine science, Guanajuato, México, 19–25 June, 2011. A12_470
9. Popov MV, Tyurin VN, Druyanov BA (1984) Counterbalanced manipulator, Patent SU1065186, January 7, 1984
10. Simionescu I, Ciupitu L (2000) The static balancing of the industrial arms. Part I: continuos balancing. Mech Mach Theory 35:1299–1311
11. Popov MV, Tyurin VN (1983) Balanced manipulator, Patent SU1000271, February 28, 1983
12. Butterfass J, Grebenstein M, Liu H, Hirzinger G (2001) DLR-hand II: next generation of a dextrous robot hand. In: Proceedings of ICRA 2001, Seoul, Korea, May 21–26 2001, pp. 109–114
13. Nathan RH (1985) A constant force generation mechanism Trans J Mech Transmissions Automation Des (107): 508–512
14. Kolotenkov VI (1984) Balanced mechanism, Patent SU1114829, September 23, 1984
15. Vrijlandt N, Herder JL (2002) Seating unit for supporting a body or part of a body, Patent NL1018178, December 3, 2002
16. Simionescu I, Ciupitu L (2000) The static balancing of the industrial arms. Part I: Discrete balancing. Mech Mach Theory 35:1287–1298
17. Minotti P, Pracht P (1988) Ressort et mécanismes: une solution aux problèmes d'équilibrage. Mech Mach Theory 23(2):157–168
18. Dzhavakhyan RP, Dzhavakhyan NP (1989) Balanced manipulator, Patent SU1521579, November 15, 1989
19. Hervé J (1985) Device for counter-balancing the forces due to gravity in a robot arm, Patent FR2565153, June 12, 1985
20. Gvozdev YF (1992) Manipulator, Patent SU1777993, November 30, 1992
21. Salisbury JK, Craig JJ (1982) Articulated hands: Force control and kinematic issues. Int J Robot Res 1(1):4–17
22. Bekey GA, Tomovic R, Zeljkovic I (1999) Control architecture for the Belgrade/USC hand in dextrous robot hands. Springer-Verlag: New York
23. Crowder RM, Whatley DR (1989) Robotic Gripping Device Having Linkage Actuated Finger Sections, U.S. Patent 4 834 443

24. Guo NS, Louis H, Meher G (1993) Surface structure and kerf geometry in AWJ: formation and optimisation. In: Proceedings of 7th American Waterjet Conference, USA, pp 1–26
25. Murai H, Nishi S (1989) Structure of waterjet and erosion of materials. In: Proceedings of 5th American waterjet conference, Canada, pp 89–98
26. Coiffet P, Gogu G (1996) Representation du mouvement des corps solides, Hermes, ISBN 2–7462-3420-3
27. Suh NP (1990) Principles of design, oxford series on advanced manufacturing, ISBN-13:978-0195043457

Part VI
Quadcopters and Mobile Robots

Active Fault-Tolerant Control of Quadrotor UAVs Based on Passive Controller Bank

Abdel-Razzak Merheb and Hassan Noura

Abstract In this paper, an Active Fault Tolerant Control (AFTC) scheme for quadrotor UAVs suffering from actuator loss of effectiveness fault is presented. The AFTC is based on Passive Fault Tolerant Controller (PFTC) bank that contains a group of Sliding Mode Controllers, each one tuned to give the best performance for a specific fault. Whenever a fault is detected and identified, the relevant PFTC is activated resulting in the fault effect kept minimum. State estimation is ensured using an Extended Kalman Filter, and fault detection and estimation process is realized using a Fault Detection and Identification (FDI) unit that examine the changes in the control signals. Simulation results using an Astec Pelican quadrotor model emphasize the effectiveness of the proposed controller in the presence of multiple actuator faults.

Keywords Fault tolerant control · Quadrotor UAVs · Sliding mode control
Extended Kalman filter

1 Introduction

Drones have started taking an important role in everyday civil life. They are covering sports competitions, filming indoor and outdoor events, monitoring wild life, and been used in several science research projects. Moreover, the decreasing cost of drones makes them an affordable high-tech game available for civilians with low pilot experience. With the increase use of small drones, accidents related to drone failures have increased. The most famous drone accident was the fall of a camera drone right behind Marcel Hirscher, a world champion skier, during a World Cup

A.-R. Merheb (✉)
International University of Beirut, Beirut, Lebanon
e-mail: abdelrazzac10@yahoo.fr

H. Noura
Islamic University of Lebanon, Khalde, Lebanon
e-mail: hassan.noura@iul.edu.lb

R. Rizk and M. Awad (eds.), *Mechanism, Machine, Robotics and Mechatronics Sciences*, Mechanisms and Machine Science 58,
https://doi.org/10.1007/978-3-319-89911-4_17

slalom race in December 2015. The accident was caught live on TV, and revealed the safety questions of drones and the importance of equipping them with fault tolerant controllers.

Switching control with multiple models was used first at the beginning of the nineties. In this scheme, many linear controllers are designed a priori to cope with different operating conditions as well as with different fault or failure types [1]. In this method, a bank of models contains models for all the possible flying conditions with their nominal controllers. The output of the system is compared to all the outputs of the models, and the controller of the model that has the nearest output is activated [2].

Many disadvantages reduce the effectiveness of this method. First, this method depends heavily on the robustness of the FDI unit. If the fault diagnosis is not accurate, the wrong controller is activated resulting in even worse response. In addition, the switching between the controllers might introduce undesired transient response that affects the overall response of the system severely. To cope for this problem, bumpless switching techniques can be used [3]. Moreover, this method uses multiple controllers designed a priori for any possible fault scenario. As a result, faults that are not predicted during the design procedure cannot be tolerated. In addition, the occurrence of multiple faults produces a new effect on the response, and all the possible fault combinations should be considered.

Authors in [4] develop a switching nonlinear controller for small quadrotors based on multiple Lyapunov functions and an adaptive mechanism. The switching between Lyapunov functions decreases the problems associated with the underactuation of the quadrotor, while the adaptive mechanism lightens the effect of the parametric uncertainty. The stability of the controller is proven analytically, and the controller is shown to be effective through simulation experiments. In [5], authors propose a switching multi-model predictive controller (MMPC) and an interacting multiple model (IMM) filter to control a realistic quadrotor model in the presence of actuator faults. Numerical simulations show good results with the quadrotor performing accurate tracking of its path despite of the fault affecting one of its rotors. A new optimal reconfiguration controller for quadrotor UAVs suffering different propeller lock and loss of control effectiveness faults is proposed in [6]. The new controller uses adaptive control and combined multiple models. Simulation results show that the asymptotic tracking of the states is ensured by appropriate switching of the multiple models, and that the convergence speed is improved by the adaptive model.

In this paper, an Active Fault Tolerant Controller based on switching control is developed for quadrotor UAVs suffering from actuator faults. The new controller uses a controller bank containing a group of Sliding Mode Controllers (SMCs), each one tuned for a specific fault.

2 The Quadrotor System

In this paper, the Asctec Pelican quadrotor model is used to develop the proposed AFTC algorithm. The state space dynamic equations of the quadrotor model shown in Fig. 1 are

Fig. 1 Quadrotor schematic

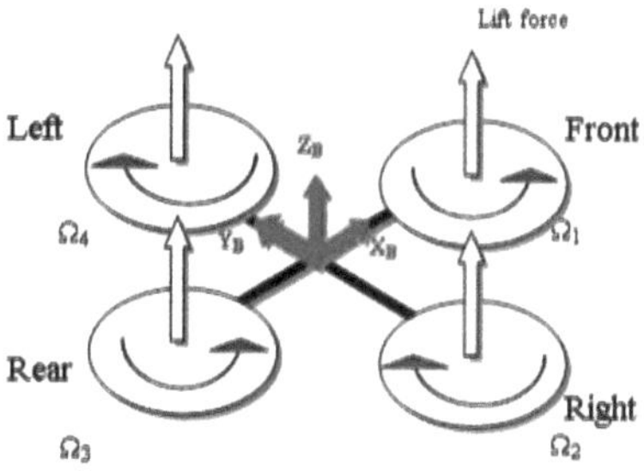

$$
\begin{aligned}
\dot{x}_1 &= x_2\\
\dot{x}_2 &= \frac{U_1}{m}(\sin x_{11}\sin x_7 + \cos x_{11}\sin x_9\cos x_7) - \frac{K_{ftx}}{m}x_2\\
\dot{x}_3 &= x_4\\
\dot{x}_4 &= \frac{U_1}{m}(-\cos x_{11}\sin x_7 + \sin x_{11}\sin x_9\cos x_7) - \frac{K_{fty}}{m}x_4\\
\dot{x}_5 &= x_6\\
\dot{x}_6 &= \frac{U_1}{m}\cos x_9\cos x_7 - \frac{K_{ftz}}{m}x_6 - g\\
\dot{x}_7 &= x_8\\
\dot{x}_8 &= \frac{I_y - I_z}{I_x}x_{10}x_{12} + \frac{I_{rotor}}{I_x}x_{10}\gamma - \frac{K_{fax}}{I_x}{x_8}^2 + \frac{U_2}{I_x}\\
\dot{x}_9 &= x_{10}\\
\dot{x}_{10} &= \frac{I_z - I_x}{I_y}x_8x_{12} - \frac{I_{rotor}}{I_y}x_8\gamma - \frac{K_{fay}}{I_y}{x_{10}}^2 + \frac{U_3}{I_y}\\
\dot{x}_{11} &= x_{12}\\
\dot{x}_{12} &= \frac{I_x - I_y}{I_z}x_8x_{10} - \frac{K_{faz}}{I_z}{x_{12}}^2 + \frac{U_4}{I_z}
\end{aligned}
\tag{1}
$$

In the equations above, the state vector is $\mathbf{x} = [x_1\ x_2\ x_3\ x_4\ x_5\ x_6\ x_7\ x_8\ x_9\ x_{10}\ x_{11}\ x_{12}\ ]^T = \left[x\ \dot{x}\ y\ \dot{y}\ z\ \dot{z}\ \phi\ \dot{\phi}\ \theta\ \dot{\theta}\ \psi\ \dot{\psi}\ \right]^T$, the control input vector is $\mathbf{u}(t) = [U_1\ U_2\ U_3\ U_4]^T$, and the output vector is $\mathbf{y} = \mathbf{h}(x) = \mathbf{C}\mathbf{x} = [x\ y\ z\ \phi\ \theta\ \psi\ ]^T$. x, y, and z are the quadrotor coordinates, ϕ, θ, and ψ are its attitude variables, and $I_{x,y,z}$ are its moments of inertia with respect to x, y, and z axes respectively. K_{ftx}, K_{fty}, and K_{ftz} are the coefficients of the translation drag forces, K_{fax}, K_{fay}, and K_{faz} are the aerodynamic friction coefficients, m is the mass of the quadrotor, and g is the gravitational constant. Table 1 shows the parameters of Asctec Pelican quadrotor used in the simulation. Note that γ is the rotor speed imbalance defined as $\gamma = Speed\ Rotor_1 - Speed\ Rotor_2 + Speed\ Rotor_3 - Speed\ Rotor_4$, and which is taken as disturbance if the rotor speeds are not measurable in flight [7].

Table 1 Parameters of the quadrotor model

I_x	$8.1e^{-3}$ N ms^2	b	$54.2e^{-6}$ N s^2
I_y	$8.1e^{-3}$ N ms^2	l	0.24 m
I_z	$14.2e^{-3}$ N ms^2	d	$1.1e^{-6}$ ms^2
I_{rotor}	$104e^{-6}$ N ms^2	g	9.8 ms^{-2}
K_{fax}	$5.56e^{-4}$ N rad/s	m	1 Kg
K_{fax}	$5.56e^{-4}$ N rad/s	K_{ftx}	$5.56e^{-4}$ N m/s
K_{faz}	$6.354e^{-4}$ N rad/s	K_{ftz}	$6.354e^{-4}$ N m/s
K_{fty}	$5.56e^{-4}$ N m/s		

2.1 Controller Scheme

The Active FTC developed here is based on Passive FTC bank. The controller bank contains several Passive Fault Tolerant Controllers, where each controller is optimized for a specific actuator fault. The Fault Detection and Identification unit is responsible for detecting the fault occurrence, finding its location, and estimating its magnitude. Whenever the fault is revealed, the convenient controller that gives the best quadrotor response in presence of the fault is activated (Fig. 2).

Fig. 2 Controller scheme

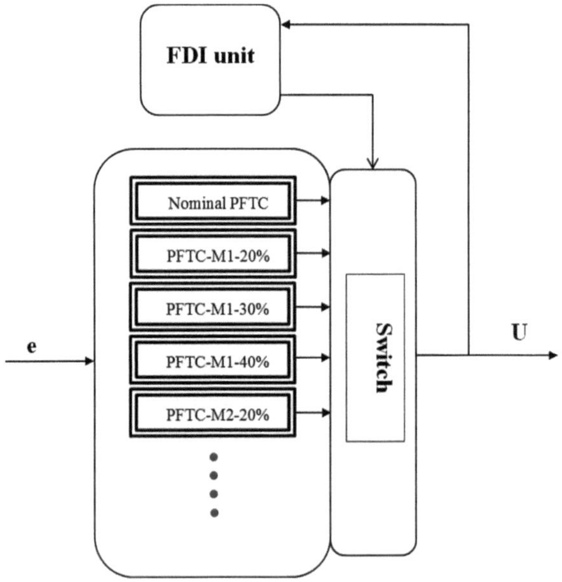

2.1.1 Passive FTC Sliding Mode Controller

The Passive FTCs in the controller bank are used to control a smaller system formed of the height of the quadrotor along with its attitude variables. The Passive FTCs are based on the following sliding mode scheme

$$\begin{cases} U_1 = \frac{m}{cos\theta cos\phi}[c_z\dot{e}_z + \ddot{z}_d + g] - k_z sat(s_z) \\ U_2 = I_x[c_\phi \dot{e}_\phi + \ddot{\phi}_d - \frac{I_y - I_z}{I_x}\dot{\theta}\dot{\psi} - \frac{I_{rotor}}{I_x}\dot{\theta}\gamma] - k_\phi sat(s_\phi) \\ U_3 = I_y[c_\theta \dot{e}_\theta + \ddot{\theta}_d - \frac{I_z - I_x}{I_y}\dot{\phi}\dot{\psi} + \frac{I_{rotor}}{I_y}\dot{\phi}\gamma] - k_\theta sat(s_\theta) \\ U_4 = I_z[c_\psi \dot{e}_\psi + \ddot{\psi}_d - \frac{I_x - I_y}{I_z}\dot{\phi}\dot{\theta}] - k_\psi sat(s_\psi) \end{cases} \tag{2}$$

where $\boldsymbol{i}_d$ is the desired state vector, and $\boldsymbol{e}$ is the state error vector ($e_i = i_d - i$) for $i = z, \phi, \theta, \psi$. s_i is the sliding surface chosen as $s_i = \dot{e}_i + c_i e_i$, c_i is the sliding surface slope with $\boldsymbol{c} = [c_z\ c_\phi\ c_\theta\ c_\psi]^t$, and $\boldsymbol{k} = [k_z\ k_\phi\ k_\theta\ k_\psi]^t$ is the discontinuous control gain. Note that $sat(\bullet)$ is the saturation function used to minimize the chattering of the discontinuous part of the controller. The discontinuous part of a sliding mode controller is responsible for maintaining its robustness in presence of modelling uncertainty, disturbance, and faults. The right choice of $\boldsymbol{c}$ and $\boldsymbol{k}$ gain vectors increases the immunity of the quadrotor system towards faults and anomalies.

2.1.2 Controller Bank

It is well known that Sliding Mode Controllers have inherent robustness against model uncertainties and disturbances. This robustness is embedded in the discontinuous gain vector $\boldsymbol{k}$ which ensures the conduction of the system states towards the sliding surface and thus, the safe control of the system.

The proposed controller tuning for $\boldsymbol{c}$ and $\boldsymbol{k}$ vectors is realized using a bioinspired search algorithm called Ecological Systems Algorithm (ESA). This algorithm starts by generating two search species, a predator and a prey species, and then iterates using the interaction between these two species until it finds an optimal solution for the problem. In our case, ESA searches iteratively for the controller gains that give the best quadrotor response. For nominal situations when no fault, no model uncertainty, and no disturbance exist, ESA uses a healthy quadrotor model to find the gains. The optimal nominal gains of the Sliding Mode Controller presented in (2) are found as $\boldsymbol{c} = [9.2267;\ 4.7203;\ 4.7306;\ 5.1215]$ and $\boldsymbol{k} = [108.0148;\ 1.0775;\ 5.2567;\ 2.6467]$. Detailed information on ESA application to find the optimal controller gains can be found in [8, 9].

By using a faulty quadrotor model to tune the controller, ESA finds the best gains that ensure high robustness of the controller in presence of the proposed fault [10]. The controller bank is formed of multiple Passive FTCs Eq. (2), each one tuned for a specific actuator fault. The faults considered in this paper are based on the partial loss of effectiveness of rotors injected in different magnitudes to different actuators.

This fault results in the decrease of the thrust delivered by a rotor, and is the result of rotor breakage after a collision, motor power problems, rotor deformation, or even debris at the motor shaft. The controller bank contains thirteen different controllers, the nominal controller, and the PFTCs tuned for 20, 30, and 40% faults injected to each rotor. Optimal gains for the quadrotor under fault are found by using a faulty quadrotor model with the amount of fault injected at the beginning of the tuning process. To find the best Passive FTC suitable for quadrotor control under 30% fault in rotor 3 for example, a quadrotor model that has a permanent fault of 30% in rotor 3 is used in the tuning process. The result of the tuning are gains that give the best performance of the quadrotor when infected with 30% loss of effectiveness fault at rotor 3. Note that because the Passive FTCs use Sliding Mode theory, any further/multiple faults are still affordable with the inherent robustness of the controllers.

2.2 *State Estimation and FDI Unit*

To estimate the states of the quadrotor online and provide feedback signals for the controller, an Extended Kalman Filter is implemented and used. The detailed Extended Kalman Filter equations used in this paper can be found in [10].

The amount and location of actuator faults are estimated by evaluating the fault effect on the control signals. Any fault affecting one of the rotors results in the quadrotor deviation from the desired path. This will drive the controller to compensate for the emerging error, and the motors will examine a sudden change in their speeds. Distinctive actuator faults show distinctive changes in the control commands, especially in U_4. Fault effect on control signals is shown in Table 2.

The FDI unit works as follows. During the quadrotor flight, U_4 control command is examined online for any sudden changes. Whenever a sudden change in U_4 is detected, control commands U_2 and U_3 are examined. If the change in the latter commands matches one of the changes in Table 2, fault alarm is given and the fault location is determined. Next, the amount of change in U_4 is measured, and Table 3 is used to estimate the fault magnitude. Table 3 is formed by injecting different faults to different motors and examining the change in U_4 offline.

Table 2 Fault effects on control commands

M1 Fault	Positive change in U_4	Negative change in U_3
M2 Fault	Negative change in U_4	Positive change in U_2
M3 Fault	Positive change in U_4	Positive change in U_3
M4 Fault	Negative change in U_4	Negative change in U_2

Table 3 Changes in U_4 for different fault magnitudes

	15%	25%	35%	45%	55%
$M1$	0.00076	0.00154	0.00271	0.00458	0.00783
$M2$	0.00076388	0.0015479	0.002727	0.0104	0.013
$M3$	0.0007618	0.001544	0.0027157	0.0045854	0.0078424
$M4$	0.00076268	0.0015462	0.0024827	7.4e-3	8.4e-3

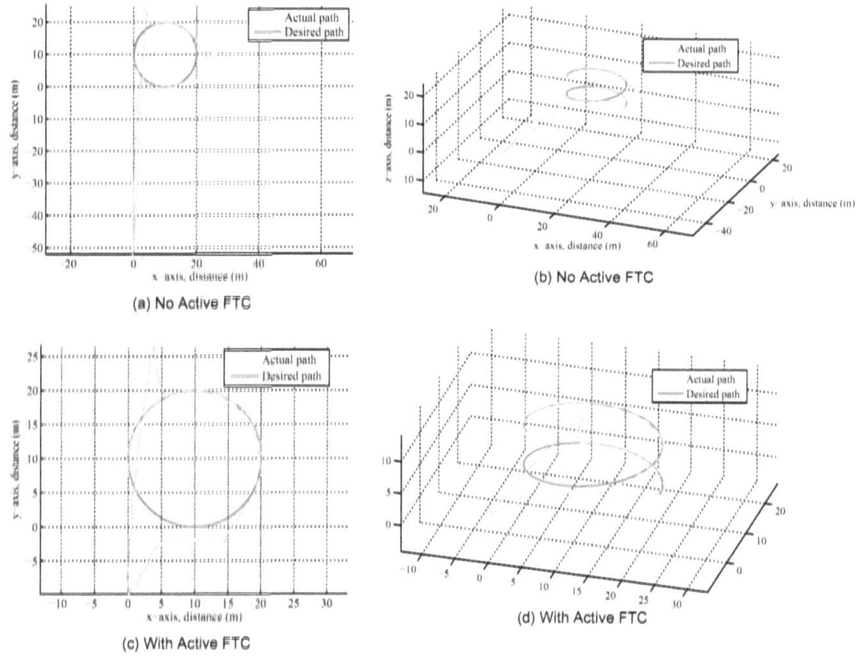

Fig. 3 Path of the quadrotor infected with fault-1

3 Simulation and Results

In this section, the efficiency of the proposed Active FTC bank is checked in presence of loss of effectiveness faults and measurement noise. The quadrotor controller has a cascaded scheme with the height and the attitude angles controlled with the Active FTC Eq. (2), and the x and y variables are controlled with a simple PD controller. The quadrotor is set to follow a 3-D path in the space with the shape of a helix, and multiple loss of effectiveness faults are injected to different motors in different instants. The first fault considered is a 30% loss of effectiveness in motor 2 at $t = 20$, followed by a 40% fault affecting motor 3 at $t = 50$. The second fault scenario is a loss of effectiveness fault of 50% injected to motor 1 at $t = 30$, followed by a 30% loss of effectiveness in motor 2 at $t = 50$. White noise is used for measurement

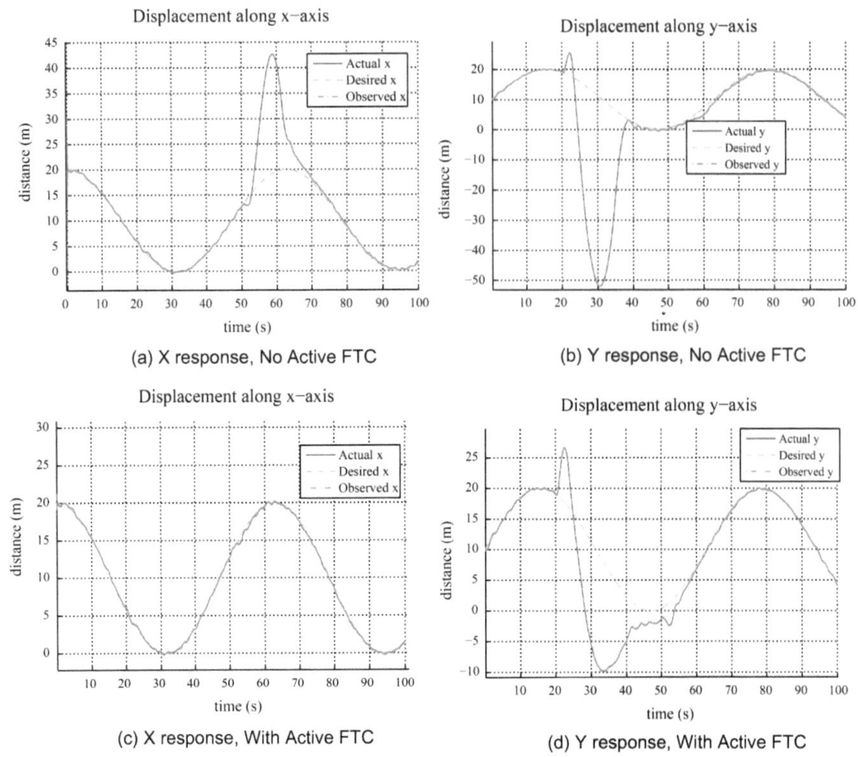

Fig. 4 Position response of the quadrotor infected with fault-1

noise, with noise amplitude of $\frac{\sqrt{noise\ power}}{3\sqrt{sample\ time}}$, and standard deviation of $\frac{\sqrt{noise\ power}}{\sqrt{sample\ time}}$. The position ($x$, y, z variables) accuracy is assumed to be 10 cm, with noise power of 10^{-6} for x and y variables, and 10^{-7} for z variable. The accuracy of the attitude angles measurement is assumed to be 0.2°, with attitude noise power of 10^{-9}. Finally, the sample time is taken as 0.001 s. The reader is directed to [11] for the 3-D path equation and detailed information about the position controller.

Simulation results show that the quadrotor controlled with the proposed controller has gain significant fault tolerant capabilities. For the first fault scenario, the quadrotor experienced a small deviation in its path despite the injection of two successive faults as shown in Fig. 3. The examination of the position variables response of the infected quadrotor reveals the effect of the proposed controller (Fig. 4). The fault in motor 2 results in a severe deviation in y variable at $t = 20$ that is not compensated by the nominal controller. A new fault injected in motor 3 affects severely x variable with no compensation effort applied by the nominal controller. When the PFTC bank is used for the same scenario, the first fault effect on y is suppressed significantly, and the second fault effect on x is almost shaded. Figure 5 shows the changes affecting the quadrotor controls infected by the first fault scenario. These changes are evaluated

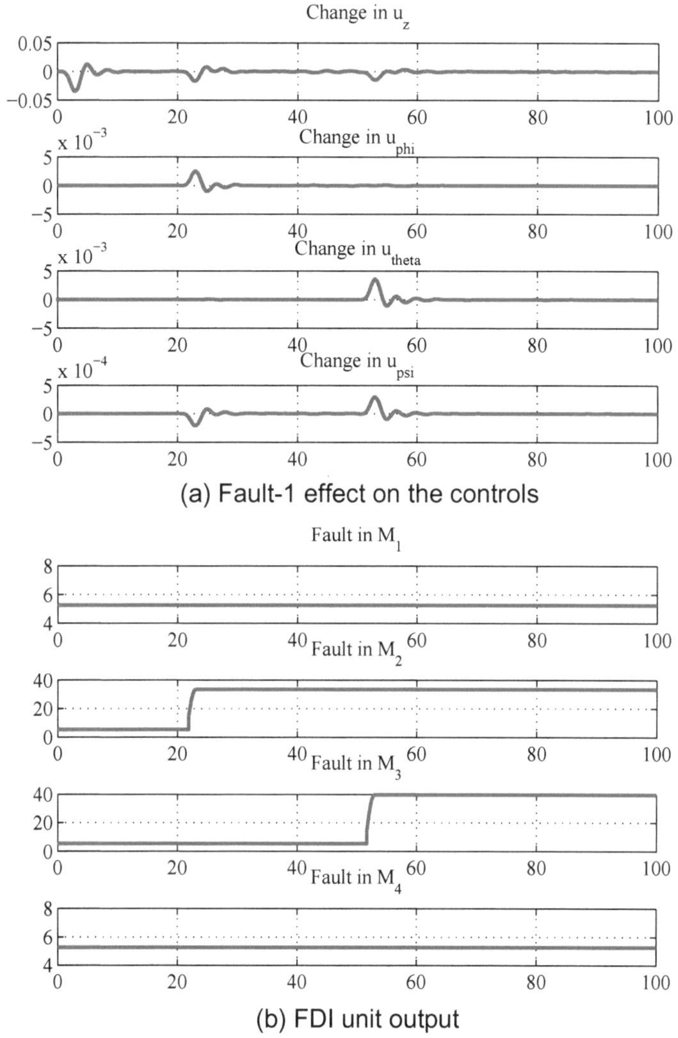

Fig. 5 Changes in the controls and output of the FDI unit with fault-1

by the FDI unit according to Tables 2 and 3 in order to estimate the fault location and magnitude online. Figure 6 shows the quadrotor path under fault scenario 2 where the severe fault is injected first. Again, the proposed controller has reduced the deviation in the quadrotor path resulting in less oscillations.

It is important to emphasize the robustness of the proposed controller in presence of multiple faults. Despite the controllers were tuned for only one fault, they are still able to handle multiple faults with good accuracy. This is because the Passive FTCs are based on Sliding Mode theory in which, a controller has inherent robust-

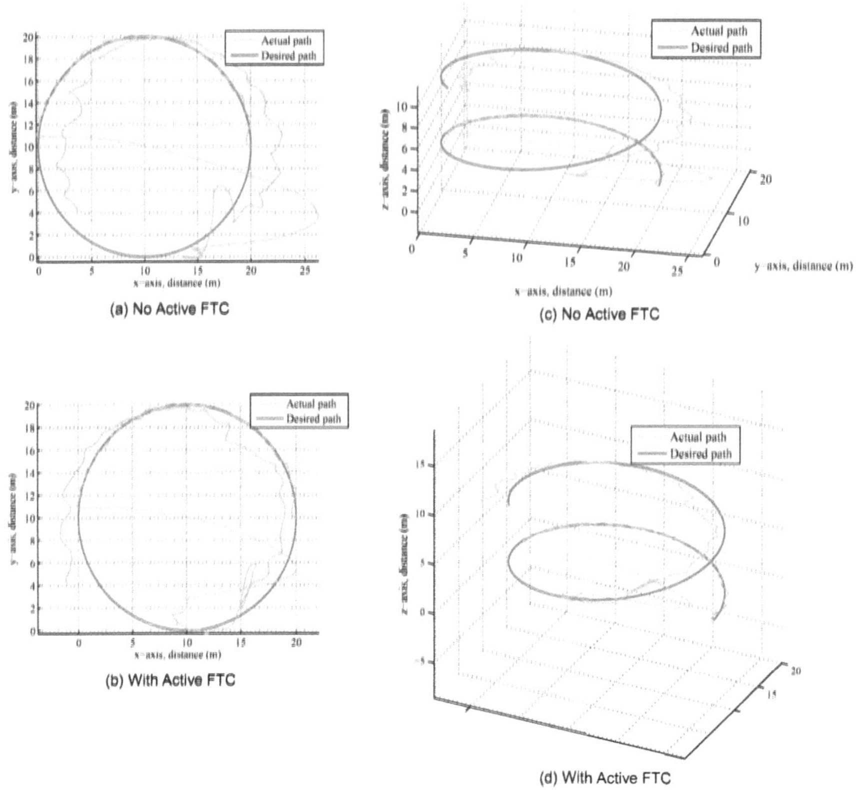

(a) No Active FTC
(c) No Active FTC
(b) With Active FTC
(d) With Active FTC

Fig. 6 Path of the quadrotor infected with fault-2

ness against disturbances and actuator faults. Taking the second fault scenario as an example, motor 1 loses half of its effectiveness at $t = 30$. The fault is detected and the relevant passive controller is activated on the fly. The result is a quadrotor system controlled with the best controller for its situation. When a new fault strikes motor 2 at $t = 50$, this motor loses 30% of its effectiveness but robustness of the Passive FTC controller is still able to handle the new fault with less effect on the overall reponse of the quadrotor.

Note that in our work, nothing is provided about the stability of the switching controller. A detailed stability analysis of the proposed controller following the work done by Villanueva et al. [12] is left for future work.

4 Conclusion and Future Works

A switching Active Fault Tolerant Controller based on Passive Fault Tolerant Controller bank is proposed for quadrotor UAVs. The new controller evaluates the changes in the control laws in order to detect and estimate loss of effectiveness faults affecting the quadrotor motors. Once the fault location and magnitude is estimated, the relevant Passive Fault tolerant Controller is activated among all controllers found in the controller bank. The controller bank includes several Passive FTCs, each one tuned with a bioinspired search algorithm to give the optimal response for a specific fault. SIMULINK results show that the new controller was successful in improving the quadrotor immunity against actuator faults, despite the occurrence of multiple actuator faults. Future work should focus on the real-time implementation of the controller on a quadrotor, and the intensive study of the stability of the system while switching its controller.

Acknowledgements This research is supported by the International University of Beirut (BIU), Lebanon.

References

1. Alwi H, Edwards C, Tan C (2011) Fault detection and fault-tolerant control using sliding modes. Springer Science & Business Media
2. Molenkamp D (2016) Intelligent controller selection for aggressive quadrotor manoeuvring, a reinforcement learning approach, Master thesis, Faculty of Aerospace Engineering, Delft University of Technology, Delft, Netherlands, April 2016
3. Qi W, Baq W, Zhang Q, Chang J (2016) Bumpless switching control for switched systems with partial actuator failures. Int J Syst Sci 47(15):3554–3560
4. Emran B, Najjaran H (2016) Switching control of quadrotor with adaptation mechanism. In: IEEE international conference on systems, man, and cybernetics (SMC), Budapest, Hungary, October 2016
5. Cândido A, Galvão R, Yoneyama T (2014) Actuator fault diagnosis and control of a quadrotor. In: 12th IEEE international conference on industrial informatics (INDIN), Porto Alegre, Brazil, July 2014
6. Chen F, Wu Q, Tao G, Jiang B (2014) A reconfiguration control scheme for a quadrotor helicopter via combined multiple models. Int J Adv Robot Syst 11(8):1–10
7. Bouabdallah S (2007) Design and control of quadrotors with application to autonomous flying, PhD thesis, École Polytechnique Fédérale De Lausanne
8. Merheb A, Noura H, Bateman F (2013) Passive fault tolerant control of quadrotor UAV using regular and cascaded sliding mode control. In: 2nd international conference on control and fault-tolerant systems, SysTol13, Nice, France, October 2013
9. Merheb A, Noura H (2012) Novel bio-inspired stochastic tuning of a quadrotor PD controller. In: The second annual Australian control conference (AUCC 2012), Australia, Sydney
10. Merheb A (2016) Diagnostic and Fault-Tolerant Control Applied to an Unmanned Aerial Vehicle, PhD thesis, cole Doctorale en Mathematiques et Informatique, Aix-Marseille University, Marseille - France, December 2016
11. Merheb A, Noura H, Bateman F, Al-Jaroodi J (2015) Fault severity based integrated fault tolerant controller for quadrotor uavs. In: 2015 international conference on unmanned aircraft systems (ICUAS'15), Denver, CO, USA, June 2015
12. Villanueva A, Castillo-Toledo B, Bayro-Corrochano E, Luque-Vega L, Gonzalez-Jimenez L (2015) Multi-mode flight sliding mode control system for a quadrotor. In: International conference on unmanned aircraft systems (ICUAS), Denver, Colorado, USA, June 2015

Low-Cost Quadcopter Indoor Positioning System Based on Image Processing and Neural Networks

I. Hatem, M. Jamal, Y. Murhij and Z. Ali

Abstract Quadcopter positioning in indoor environments is considered a major problem because of the difficulty of estimating a reliable position. Moreover, the positioning system is expected to work in real-time and to be accurate and cost-effective. In this paper, a combination of image processing techniques and neural networks is proposed to obtain the quadcopter position along the X, Y and Z coordinates. Three neural networks were used, one for each dimension. The proposed neural network based technique estimates the quadcopter target position along X, Y, and Z from two image points extracted from images captured by two low-cost IP cameras. The offered positioning system has been implemented on a locally designed and assembled quadcopter. Hovering experiments on the quadcopter have been performed in an indoor lab based environment. The results show that combining image processing techniques with neural network-based method achieves a low-cost accurate positioning system within a precision of a few centimeters with a frequency of 16 Hz.

Keywords Image processing · Indoor quadcopter positioning · Neural networks

I. Hatem (✉) · M. Jamal · Y. Murhij · Z. Ali
Tishreen University, Latakia, Syria
e-mail: iyadhatem@tishreen.edu.sy

M. Jamal
e-mail: mayssjamal@gmail.com

Y. Murhij
e-mail: yosha.morheg@gmail.com

Z. Ali
e-mail: zainalabedeenali@gmail.com

R. Rizk and M. Awad (eds.), *Mechanism, Machine, Robotics and Mechatronics Sciences*, Mechanisms and Machine Science 58,
https://doi.org/10.1007/978-3-319-89911-4_18

1 Introduction

Currently, the most commonly used positioning technique in quadcopters is the global navigation satellite systems (GNSS). However, global navigation satellite systems are generally not suitable to establish indoor locations since microwaves are attenuated and scattered by roofs, walls and similar barriers [1, 2]. In order to make positioning data available, indoor positioning system can be used. Most common indoor positioning systems are wireless-based, laser-based or vision-based technologies. Wireless and laser based technologies provide sufficient accuracy and enough low latency for stable vehicle control but they need further development work. Passive systems like motion capture systems are very popular choices for indoor environments. They can provide increased accuracy at the expense of overpriced equipment and installations. However, technologies used in indoor positioning system depend on the type of application. For an autonomous quadcopter project, as in our application, a positioning system is needed to track the location of the quadcopter accurately inside a lab area. This system need to be real-time, cost-effective and have a precision of a few centimeters for the X, Y and Z axes. To accomplish this, we implemented a real-time 3D positioning system based on visual features and three trained neural networks using two low-cost IP cameras.

2 Related Work

Previous work on quadcopter positioning can be categorized into three different research areas. One area focuses on vision technology such as stereo vision, another area focuses on wireless technologies and Ultra-wideband range measurements, and the rest focuses on monocular Simultaneous Localization and Mapping (SLAM). In 2012, a stereo vision sensor was introduced as an indoor positioning system for UAVs by Mustafah et al. [3]. The system utilizes two video cameras for stereo vision and a set of fast algorithms to obtain position information in real-time. The conducted experiment showed that the system could provide a reliable accuracy in real-time. In 2016, Guo et al. [4] introduced a localization system for quadcopters by using measurements from ultra-wideband range. In this system, an ultra-wideband module on the quadcopter communicates with fixed modules at known positions to obtain a distance. This distance is fed to a localization algorithm after calibration and outlier detection process. Extended Kalman filter (EKF) was used in this algorithm to sustain initialized trilateration. Recently, a localization using a monocular SLAM framework was introduced by Shree et al. (2017) [5]. Indoor localization and mapping new areas were combined and run together in a framework of SLAM. Onboard cameras and a cascaded position controller along with a Luenberger observer (which can combine

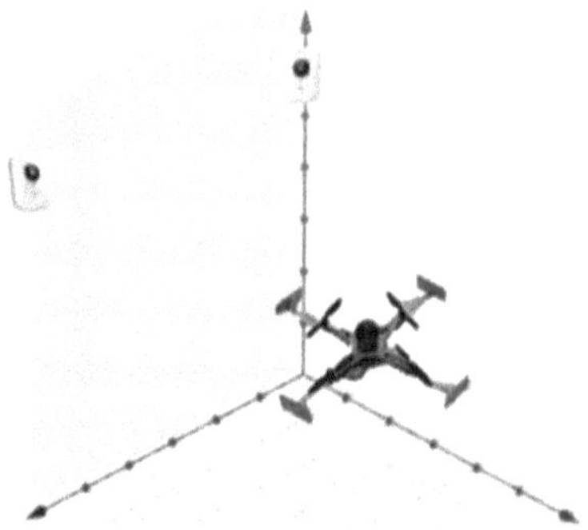

Fig. 1 3D representation of the positioning system

the data of inertial sensors and vision based position to generate a complete velocity feedback for the system) have been used. Sensor data fusion using EKF have been performed for scale estimation.

3 Materials and Methods

The proposed positioning system consisted of two IP cameras of type TP-Link connected to a Core-i3 notebook computer. The quadcopter dimensions were 40 * 40 cm. The IP cameras were positioned on a wall on height of 2.8 m: one in the corner horizontally aligned with the other camera on a distance of 4 m (Fig. 1). The work was implemented by using C++ on CodeBlocks IDE, Opencv image processing library and MATLAB.

4 Detecting the Object of Interest

Because of its distinctive color and invariant-direction shape, the colored ball on the quadcopter easily allowed the system to detect its position. First, video frames were grabbed from each camera and converted from RGB color model into HSV color model (Figs. 2 and 3). To reduce the variation in edge pixels due to light density changes, a Gaussian filter with a suitable mask was used to blur the resulting image and enhance the edges of the target object. Next, the resulting image was thresholded to segment the object of interest (the red ball). To eliminate unwanted small areas in the resulting image, morphological operations were applied including erosion, dilation and connected components [6] (Figs. 4 and 5). The object coordinates in

Fig. 2 Original camera frame

Fig. 3 Frame after converting to HSV

the image were determined by the center of mass of the segmented image through Eq. (1).

$$\{x, y\} = \{M_{10}/M_{00}, M_{01}/M_{00}\} \tag{1}$$

$$\text{where}: \quad M_{ij} = \sum_x \sum_y \left(I\,(x, y)\,.x^i.y^i\right) \tag{2}$$

where $I\,(x, y)$ is pixel intensity at position x, y.

Fig. 4 Segmented image

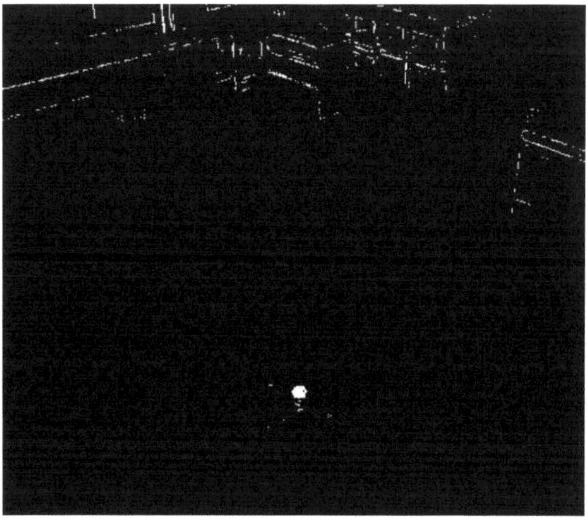

Fig. 5 Final image after morphological operations

5 Building a 3D Positioning Neural Networks System Using Information from Two IP Cameras

To build the positioning system, three neural networks were developed, for X, Y and Z. Each network has four inputs: x_1, y_1, x_2, y_2 which are the pixel coordinates of object center in the image frames grabbed from the first and the second camera, respectively. Every neural network was a feed-forward network and had one hidden

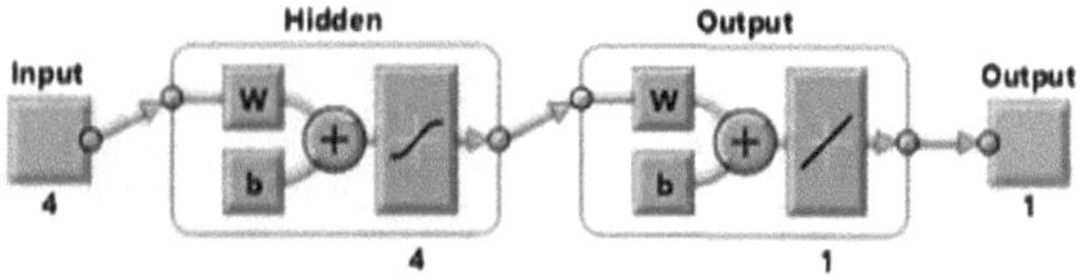

Fig. 6 Neural network structure (Matlab)

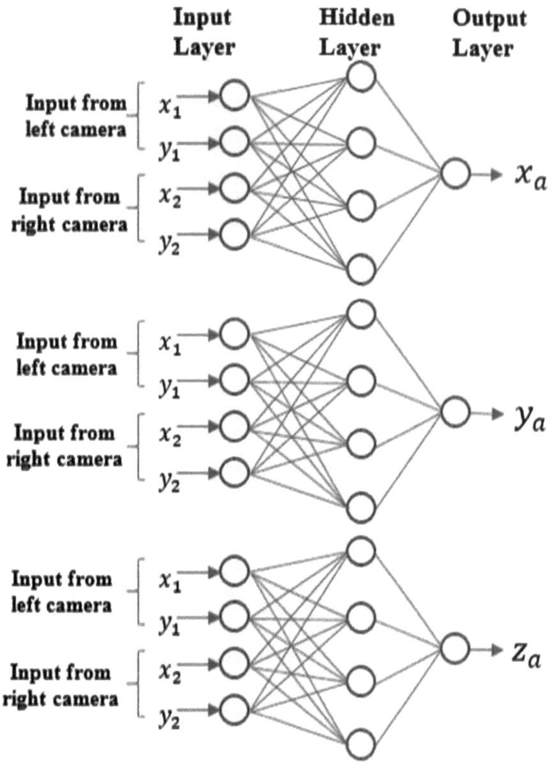

Fig. 7 System structure

layer. The hidden layer had four neurons. See Figs. 6 and 7. Neurons used in the hidden layer had *Tan-sig* activation function with the Eq. (3).

$$a = 2/\left(1 + e^{-2n}\right) - 1; \tag{3}$$

where n is the input of the neuron.

Each network had one output: the estimated position along one of the three Cartesian axes X, Y and Z in our virtual 3D system. Networks had been trained using *Levenberg Marquardt Algorithm* [7, 8]. More than one hundred distributed data samples were taken manually by measuring the real distance of the object along a virtual 3D space constructed in the workspace.

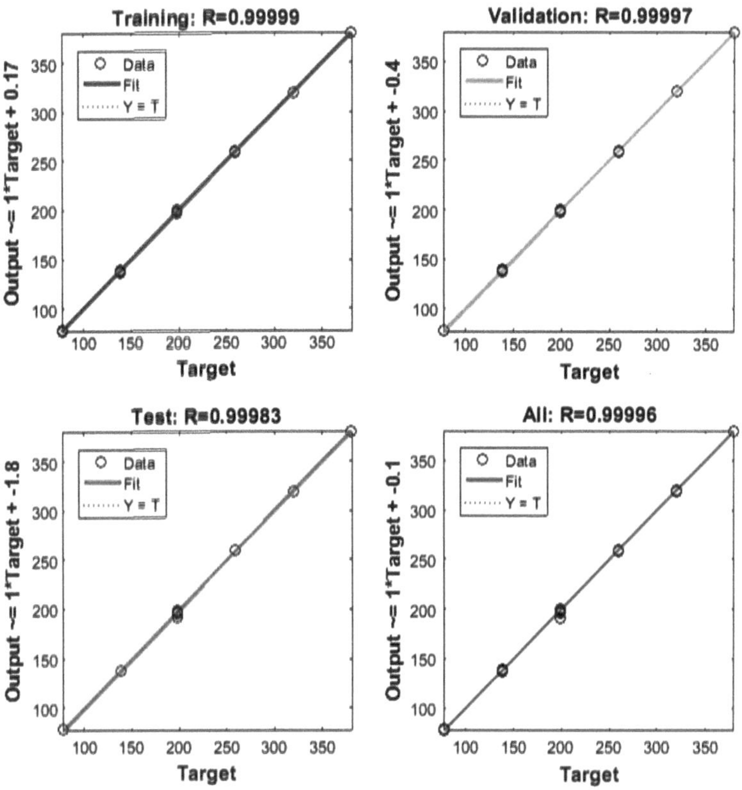

Fig. 8 Regression related to X axis neural network

Neural networks were trained on 80% of samples and validated on 10% samples and tested on the remaining 10% of the samples. Matlab software was used to construct and train the neural networks based system to get the appropriate weights and offsets for each neuron. The training regressions are shown in Figs. 8, 9 and 10. The positioning system performances on X, Y and Z axes on the training data are shown in Figs. (11-a, 12-a, 13-a) which show the current and set position of the quadcopter. Figures (11-b, 12-b, 13-b) show the error in position in tested data.

6 Implementation and Results

Testing this positioning system in a real-time application of quadcopter on new data sets showed a similar precision to the training results and proved the validity of using this indoor positioning method. Figure 14 shows the real quadcopter trajectory versus the output of the positioning system. Figures 15, 16 and 17 represent

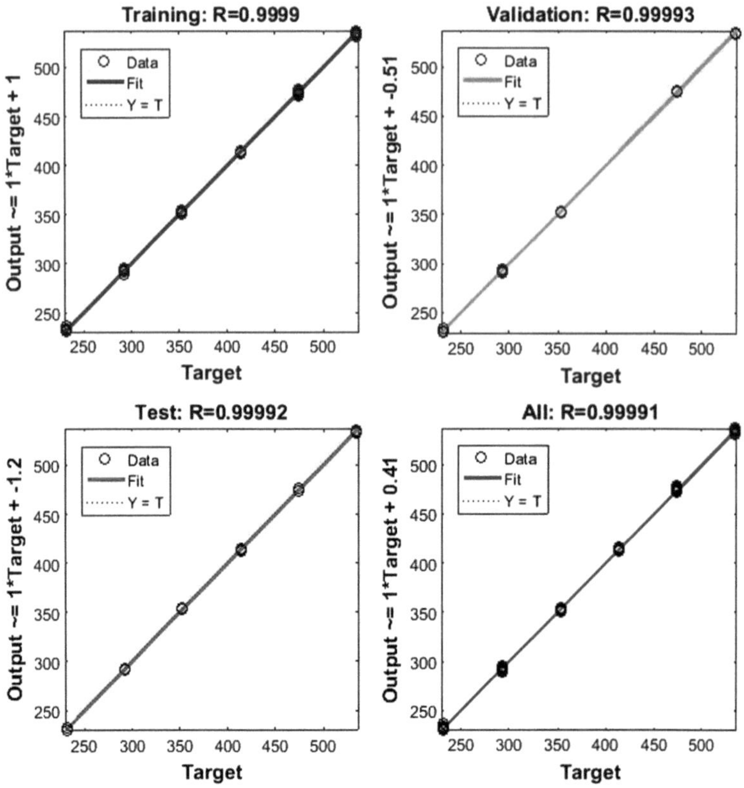

Fig. 9 Regression related to Y axis neural network

projections on X-Y plan [X = constant], X-Y plan [Y=constant] and Y-Z plan, respectively. Results show that the system has approximate precisions of ±1 cm on X axis positioning neural network and ±4 cm on Y axis positioning neural network and ±3 cm on Z axis positioning neural network with a frequency of 16 Hz. Flying tests have been applied and showed that these specifications were satisfying to control the quadcopter in real time.

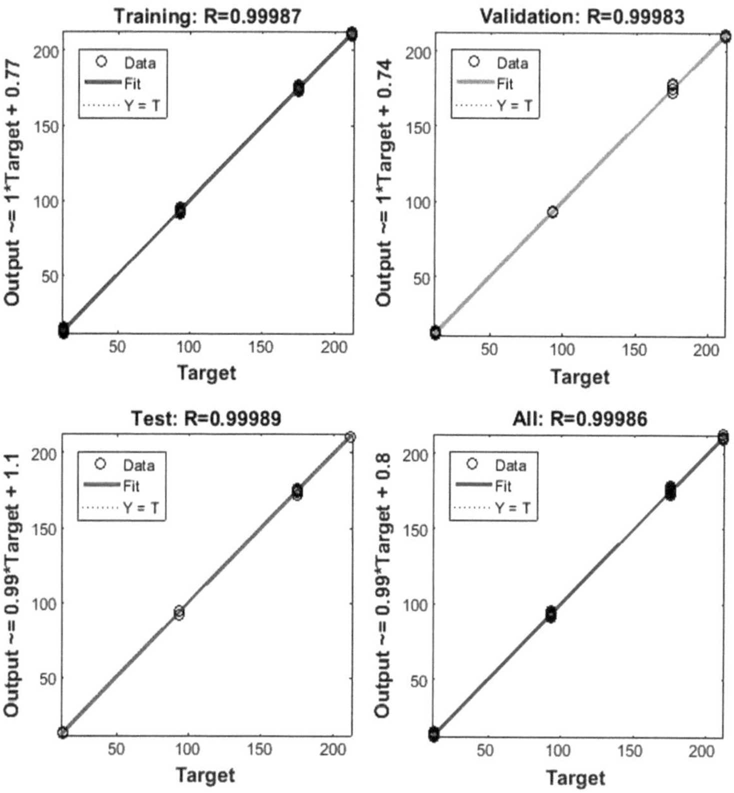

Fig. 10 Regression related to Z axis neural network

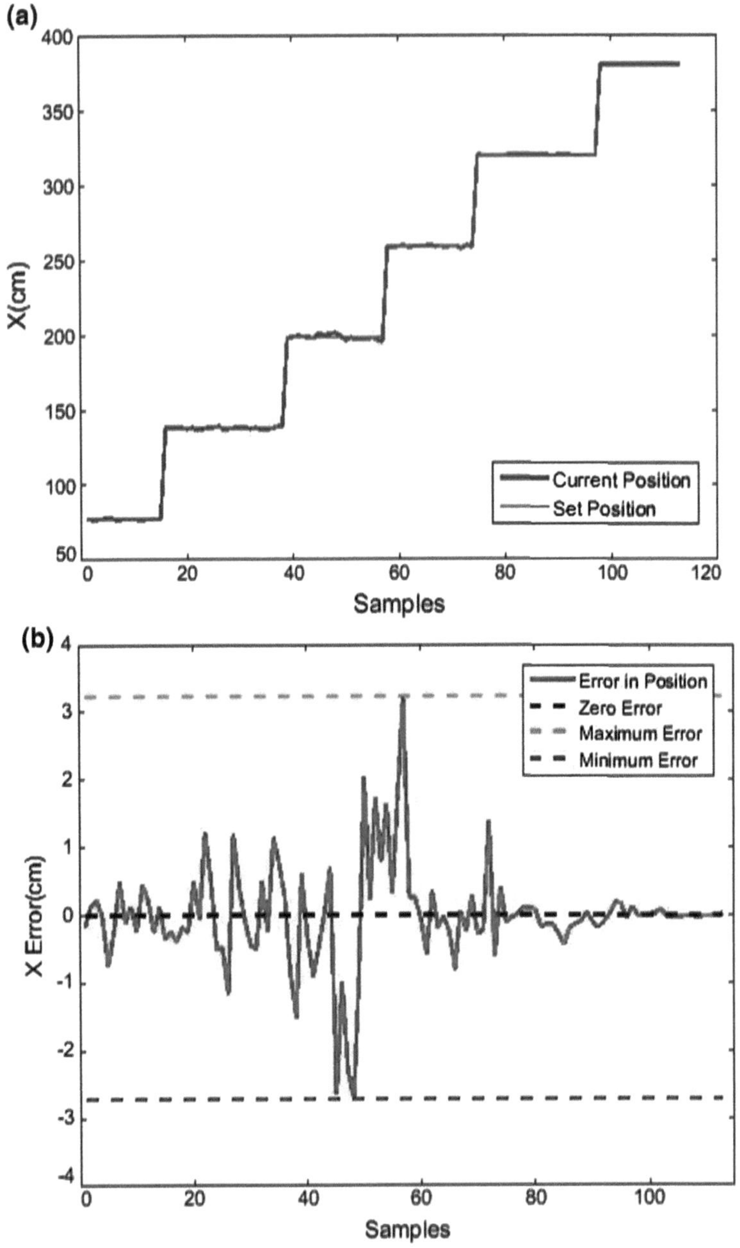

Fig. 11 **a** Current position versus set position along X-axis (cm)—**b** Error in position along X-axis (cm)

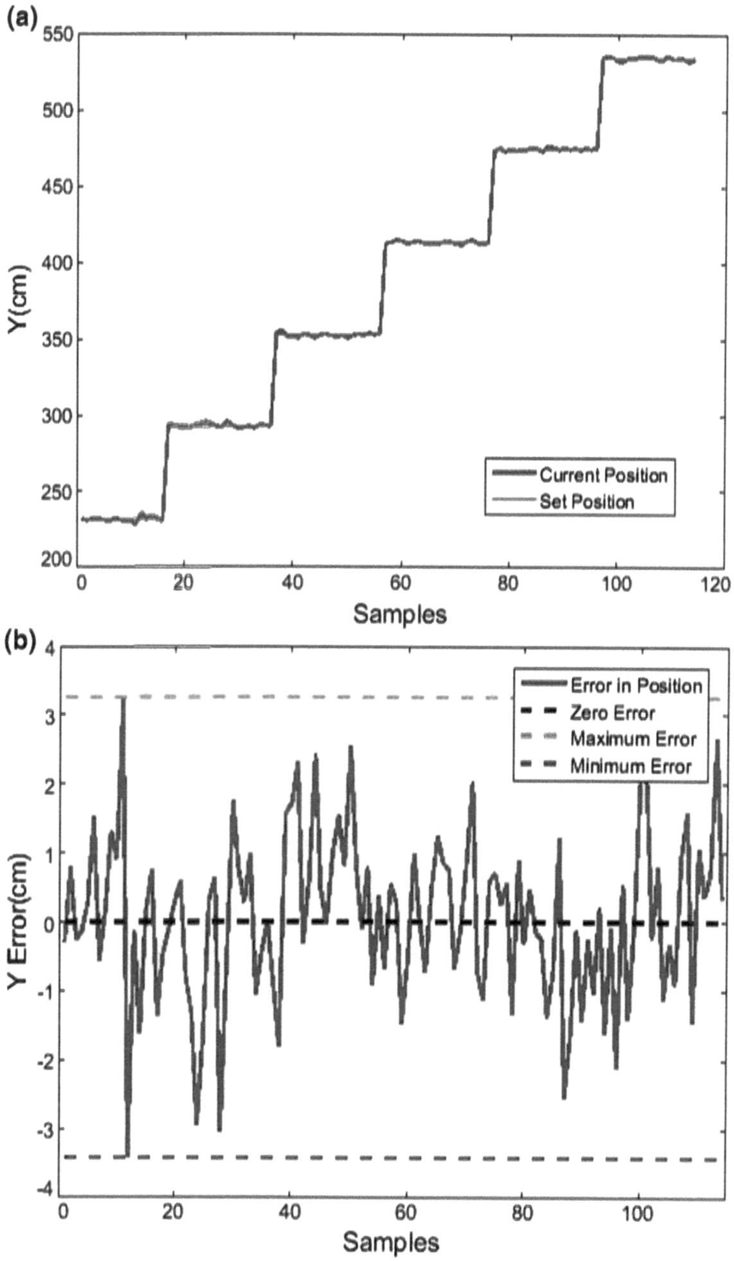

Fig. 12 **a** Current position versus set position along Y-axis (cm)—**b** Error in position along Y-axis (cm)

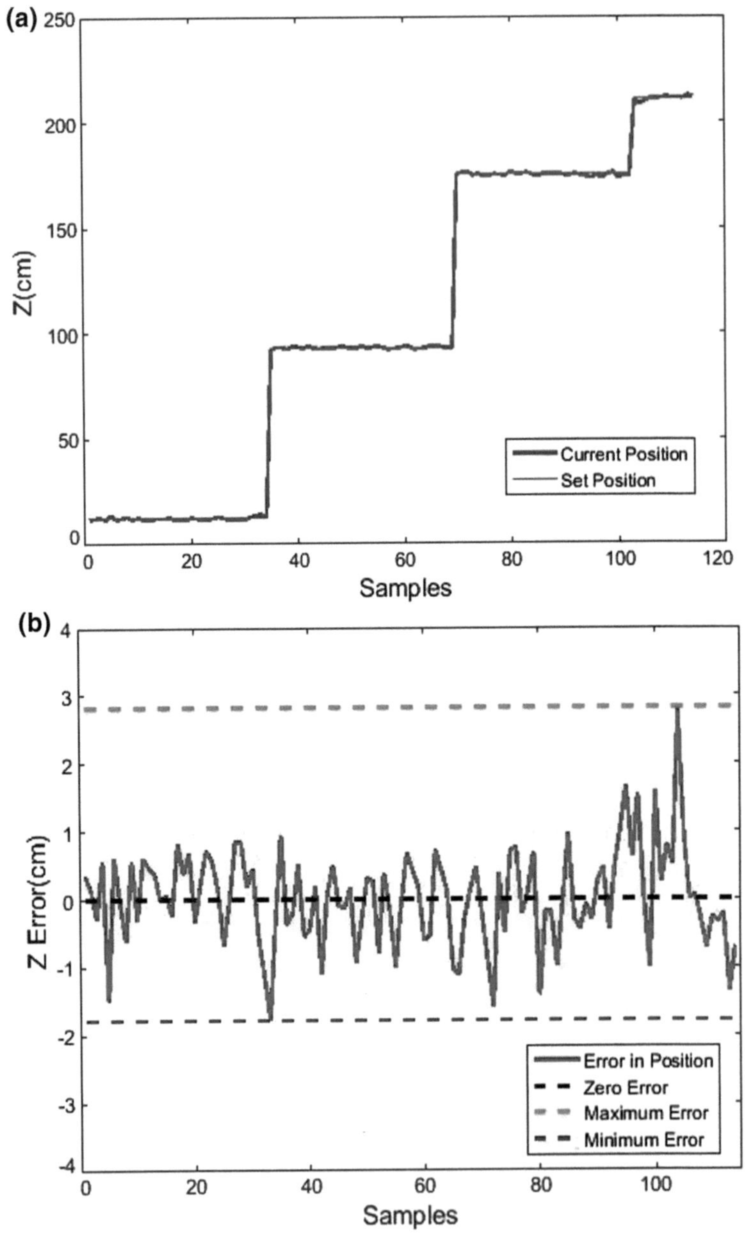

Fig. 13 **a** Current position versus set position along Z-axis (cm)—**b** Error in position along Z-axis (cm)

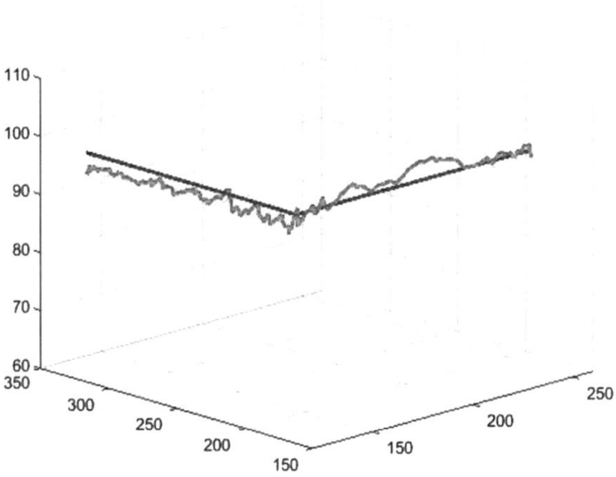

Fig. 14 Simple 3D flight trajectory

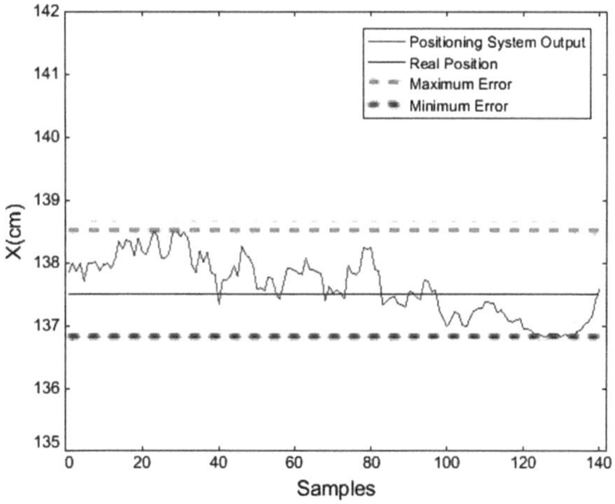

Fig. 15 X-Y plan projection [X = const]

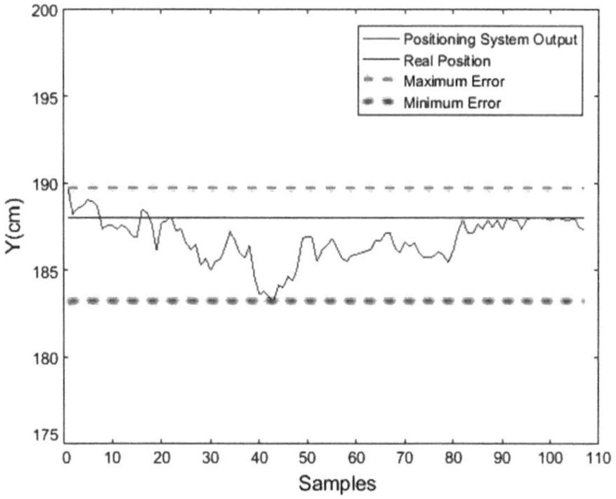

Fig. 16 X-Y plan projection [Y = const]

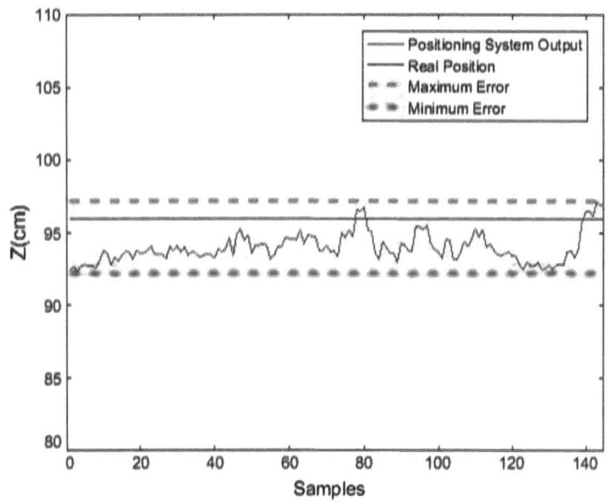

Fig. 17 Y-Z plan projection

7 Conclusions

This study has demonstrated the validity of the proposed neural-network and image-processing based technique to estimate the real-time 3D position of a flying quadcopter. First, image processing techniques to detect the object were introduced. Then, the network structure, network training and training results were explained in detail. Finally, the positioning system performances on X, Y and Z axes for a flight test of the quadcopter were presented. Results demonstrates that the proposed positioning system is a low-cost good-precision positioning system with a precision of few centimeters capable of controlling the position of the quadcopter in real time.

Acknowledgements This research was supported by the Program of Mechatronics for Distinguished at Tishreen University, Syria. The program is affiliated with the Syrian Commission for Distinction and Creativity.

References

1. Bejuri WMYW, Mohamad MM, Zahilah R, Radzi RM (2015) Emergency rescue localization (ERL) using GPS, wireless LAN and camera. Int J Softw Eng Appl 9, 217–232
2. Horvath Z, Horvath H (2014) The measurement preciseness of the GPS built in smartphones and tablets. Int J Electronics Commun Technol 1, 17–19
3. Mustafah YM, Azman AW, Akbar F (2012) Indoor UAV positioning using stereo vision sensor. Elsevier Ltd
4. Guo K, Qiu Z, Miao C, Zaini AH, Chen CL, Meng W, Xie L (2016) Ultra-Wideband-Based localization for quadcopter navigation. Unmanned Syst 4(1):23–34
5. Shree AS, Sharma RS, Behera L, Venkatesh KS (2017) Position based visual control of the hovering quadcopter. Springer, Switzerland, pp 15–26
6. Gonzalez R, Woods R (2008) Digital image processing. Prentice Hall

7. Argonne National Lab (1977) United States. Department of Energy. Levenberg-Marquardt Algorithm: Implementation and Theory
8. Hagan MT, Demuth HB, Beale MH, De Jess O (2014) Neural network design, 2nd edn. Springer, Heidelberg

UKF-Assisted SLAM for 4WDDMR Localization and Mapping

Abdulkader Joukhadar, Dalia Kass Hanna, Andreas Müller and Christoph Stöger

Abstract Correct mobile robot localization requires precise knowledge of the robot's pose in plane, i.e. the Cartesian x and y coordinates and yaw angle θ. Mobile robot pose information estimated from on-board odmetry sensors is not fully trusted and it suffers from unceratinties exerted by the robot incorporated with actuators non-linearities and robot mechanical complexities which lead to a low degree of believe (DoB) of the robot localization. The present paper provides Unscented Kalman Filter (UKF) based approach assisted robot localization to provide trusted information with high DoB for the mobile robot's pose. Particularly, estimating the current situation of the robot navigation system is complex due to the above mentioned phenomenons. An efficient and accurate estimation technique which applies probabilistic algorithm based UKF is proposed. The proposed technique is implemented and verified using MATLAB/SIMULINK®. Both practical and simulation results have demonstrated the vitality of the proposed estimation approach.

Keywords Mobile robot · UKF · SLAM · Optimal control

1st ICAMMRMS, Beirut, Lebanon, 17–19 October, 2017.

A. Joukhadar (✉) · D. K. Hanna
University of Aleppo, Aleppo, Syria
e-mail: ajoukhadar@alepuniv.edu.sy

D. K. Hanna
e-mail: eng.dalia.hanna@gmail.com

A. Müller · C. Stöger
Johannes Kepler University, Linz, Austria
e-mail: a.mueller@jku.at

C. Stöger
e-mail: christoph.stoeger@jku.at

R. Rizk and M. Awad (eds.), *Mechanism, Machine, Robotics and Mechatronics Sciences*, Mechanisms and Machine Science 58,
https://doi.org/10.1007/978-3-319-89911-4_19

1 Introduction

For autonomous mobile robots, the capability to make distinctive observation of the environment is essential to ensure a high reliability navigation performance [1].

Mobile robot pose information estimated from relative pose measurements provided by its onboard sensors is not fully trusted, and it suffers from high uncertainties which result in low degree of belief (DoB) of the robot Localization [2, 3]. To solve this problem Absolute Measurement system needs to be fused with onboard measurements using probabilistic inference algorithm [4]. Nonlinear estimation techniques play an important role in any autonomous navigation system. The fact is, serious of robot localization problems would be encountered to realize a highly trusted autonomous mobile robot [5].

For recent localization techniques, the existing algorithm are based on laser range Finder (LRF), Sonar sensors and artificial Landmarks such as barcode and visual patterns [6, 7]. Vision based mobile robot localization approaches, which use stable natural landmarks in unmodified environments are highly desirable for wide range of applications [8–10].

The Landmarks are detected by matching the key points in the image with the landmarks in database using scale invariant feature transformer (SIFT) [3].

As the image features are not noise-free, which leads to an error for the landmark and the robot's pose, Kalman Filter (KF) is used to track the landmark in dynamic environment [11]. Unscented Kalman Filter (UKF) can be applied in Simultaneous Localization and Mapping with vision system for more accuracy [12].

A novel approach presented by [1] for landmark selection, which aims at selecting the minimal set of landmarks that bounds the uncertainty in the deviation of the robot from its desired trajectory. The main goal of this paper is to develop and test an Unscented Kalman Filter (UKF) based approach-assisted absolute pose measurements (2D landmaks map) for robot localization to track the robot pose of a 4WDDMR as it moves with high DoB. To validate the proposed approach, experiments are performed using MATLAB/SIMULINK model of a 4WDDMR augmented with UKF-based algorithm developed using C code. Simulation results exhibited the effectiveness and the accuracy of the proposed algorithm compared to EKF localization based 2D landmarks map. The proposed UKF based approach has shown its inherited advantages in handling the 4WDDMR with high nonlinearities and superiority to EKF approach in avoiding the bias estimation errors caused by EKF linearization process.

The remaining sections of the paper are as follows: Sect. 2 provides details about the landmarks localization system and how the robot estimates its pose based on the relation between the landmarks and the Global coordinate frame; the localization problem solution using UKF is briefly recalled in Sect. 3; Sect. 4 reports efficient MATLAB/SIMULINK results for UKF performance with high uncertainty in the system and environment; discussion and conclusion are drawn in Sect. 5.

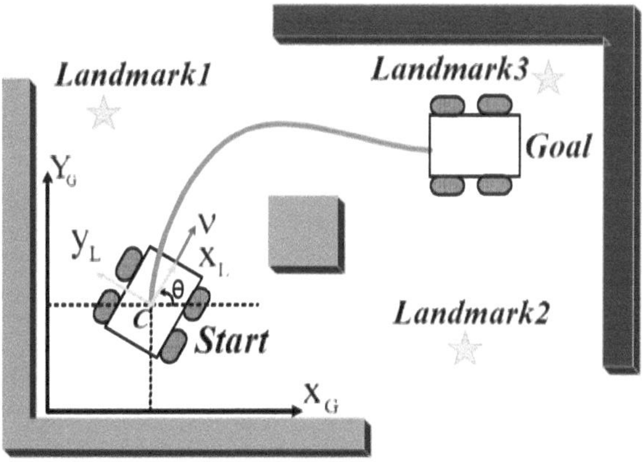

Fig. 1 Landmarks Based 4WDDMR localization system

2 4WDDMR Localization Based 2D-Landmark Map

2D-Landmarks are distinct feature that a Four Wheeled Differential Drive Mobile Robot (4WDDMR) can recognize from its sensory input and keep it to localize itself geometrically [13]. Landmarks for a certain environment are either given priori or learned by the robot. If a robot carries only proprioceptive sensors that monitor its relative motion, then the position of the robot is not accurately observable due to the accumulated errors in the robot position calculation [14]. This may lead to an expected failure in pose estimation when implementing EKF approach due to linearization problem and associated accumulation errors. Hence, the EKF estimate will drift away from a precise pose estimate unless absolute position information is provided [15]. In order to update more accurate position estimate and thus keep the robot uncertainty bounded, most of the current research efforts incorporate some form of exteroceptive information e.g., landmarks, vision system, laser scanner and GPS, which are basically based on the nature of the robot for indoor or outdoor applications [13]. Consider the situation shown in Fig. 1 a 4-WDDMR acts in the predefined environment in the global frame {G}. The landmarks predefined in the robot environment would help the robot to localize correctly in the final destination, which represents the robot final goal.

The time discrete kinematic model for the 4WDDMR with sample time T_s is given by (1):

$$\mathbf{x}_k \equiv f(\mathbf{x}_{k-1}, \mathbf{u}_k) = \begin{bmatrix} x_{k-1} + \nu_k T_s \cos(\theta_{k-1}) \\ y_{k-1} + \nu_k T_s \sin(\theta_{k-1}) \\ \theta_{k-1} + \omega_k T_s \end{bmatrix} \tag{1}$$

where $f(\mathbf{x}_{k-1}, \mathbf{u}_k)$ is a nonlinear function, relates the 4WDDMR's pose $\begin{bmatrix} x_k \; y_k \; \theta_k \end{bmatrix}^T$ in the global frame {G}, with the input vector $\mathbf{u}_k = \begin{bmatrix} \nu_k \; \omega_k \end{bmatrix}^T$ the 4WDDMR's

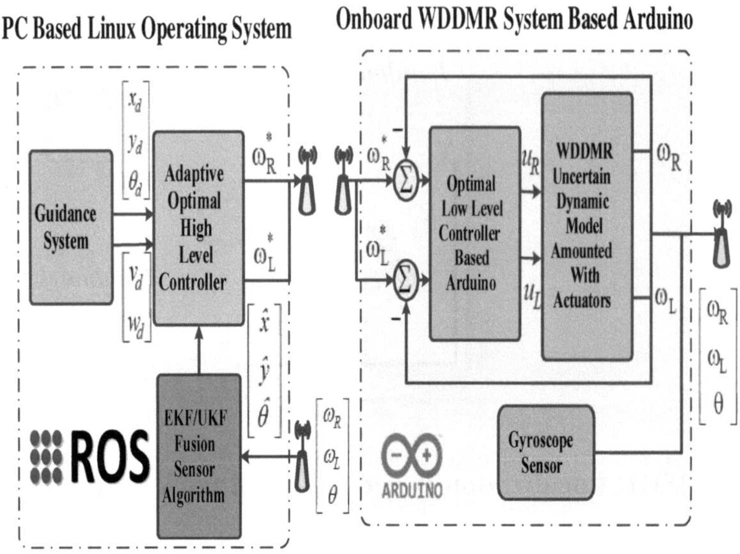

Fig. 2 Schematic of ROS for 4WDDMR [17]

translation and angular velocities in the robot frame {L}. The priori knowledge about the map is represented by the location of each landmark in the global frame {G}. However, in practice, uncertainty and noise are prone in the kinematic model of the mobile robot [16]. Assuming the noise is non-additive Gaussian distributed w_k, the noisy measurement model is written as follow

$$\mathbf{x}_k = f(\mathbf{x}_{k-1}, \mathbf{u}_k, \mathbf{w}_k) \tag{2}$$

where $\mathbf{w}_k \sim N(0, \mathbf{Q})$, $\mathbf{Q}$ is the system noise covariance matrix.

The main task is then to recognize a landmark in the robot working environment with a high degree of belief (DoB), upon which to correct the robot localization with respect to the exact detected landmark. For this reason, good odometry accuracy is a prerequisite for successful landmark detection [16].

In the authors' previous work, the 4WDDMR motion is controlled by the wheels velocities to track a pre-calculated path using an Adaptive-Optimal Controller. The proposed implemented controller is shown capable to overcome the robot dynamic parameter uncertainties and robust against parameter mismatch [17]. Figure 2 shows a block diagram of the proposed mobile robot control system. Robot Operating System (ROS) and Arduino-based microcontroller board are used for the system implementation [17].

For the present work, it is assumed that the landmarks have a fixed and known position (m_x, m_y) relative to the global frame {G}, in which the robot can localize itself. Basically, a detection of a landmark can be accomplished using several methods e.g., Scale Invariant Feature Transform (SIFT) [3]. Matching between the detected

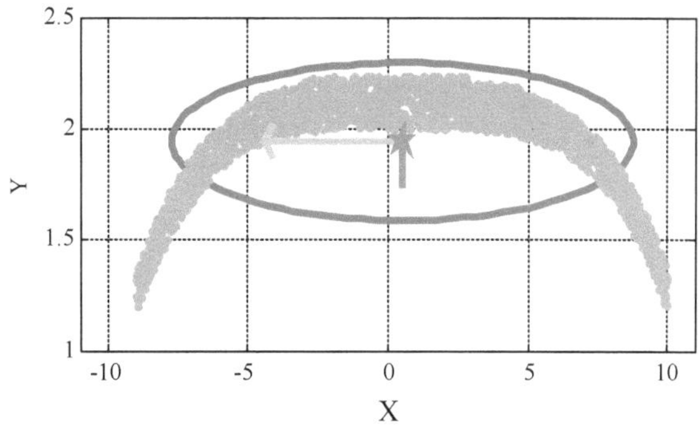

Fig. 3 Measurement distribution approximation using EKF

point and all desired landmarks stored in the system database. When matching exists, hence the image coordinates is used for the detected landmark [18]. The measurement model of location z_k of a landmark m_k from the viewpoint of the robot given the location of the robot x_k is defined as follow (3)

$$\begin{aligned}\mathbf{z}_k = h(\mathbf{x}_k, \mathbf{m}_k) &= \begin{bmatrix} h_x(\mathbf{x}_k, \mathbf{m}_k) \\ h_y(\mathbf{x}_k, \mathbf{m}_k) \end{bmatrix} \\ &= \begin{bmatrix} (m_{xk} - x_k)\cos(\theta_k) + (m_{yk} - y_k)\sin(\theta_k) \\ -(m_{xk} - x_k)\sin(\theta_k) + (m_{yk} - y_k)\cos(\theta_k) \end{bmatrix}\end{aligned} \tag{3}$$

where $h(\mathbf{x}_k, \mathbf{m}_k)$ is the measurement function that relates the robot and the landmark locations to measurement. However, in practice, there may be uncertainty and noise in landmark detection system. Assuming that this noise is Gaussian distributed $\mathbf{v}_k$, the noisy measurement model is written as (4):

$$\mathbf{z}_k = h(\mathbf{x}_k, \mathbf{m}_k, \mathbf{v}_k) \tag{4}$$

where $\mathbf{v}_k \sim N(0, \mathbf{R})$, $\mathbf{R}$ is the measurement noise covariance matrix.

Since the measured $\mathbf{x}$ and $\mathbf{y}$ coordinates have a nonlinear relationship with the orientation of the robot angle θ, the distribution of the measurements is not Gaussian. Hence, the distribution of the robot pose variables (x, y, θ) with growing uncertainties is referred as Banana Shaped as shown in Fig. 3. Thus, Gaussian distribution of these variables is no more valid [19]. Linearizing this nonlinear observation model using EKF may result in inaccuracy in the robot pose estimate, which also leads to a large bias in the mean and covariance estimation [5]. To improve the robot pose estimation accuracy, robustness and efficiency of Gaussian approximate inference algorithm applied to general nonlinear system, Unscented Kalman Filter (UKF) based approach is proposed as a fusion sensors algorithm for system localization [20].

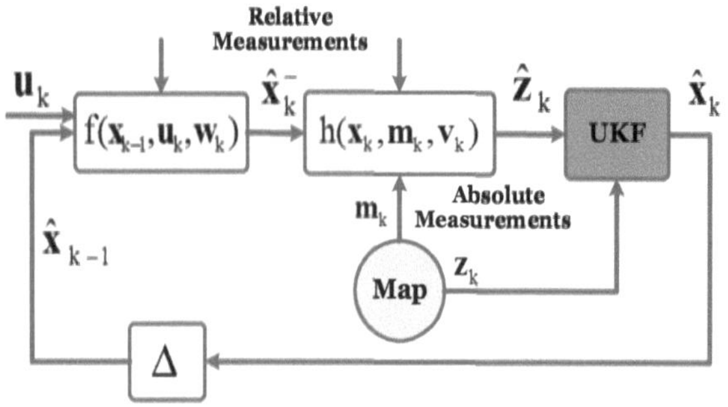

Fig. 4 Block diagram for UKF- localization

3 Unscented Kalman Filter Localization Algorithm

The present section, a UKF—localization algorithm has been investigated and verified with the presence of a large uncertainty.

In literature, researchers have developed a variety of localization approach based Probabilistic Inference depending on the 4WDDMR's action and perception system [8–10]. UKF is one of the optimum recursive filters based on Gaussian approximate solution to solve the Probabilistic Inference practically [5].

The advantage of the UKF over EKF is that UKF is not based on local linear approximation, but uses a bit further points in approximating the nonlinearity [21, 22]. The unscented transformation (UT) is able to capture the higher order moments caused by the nonlinear transform better than the Taylor series based approximation [12, 23]. In order to apply the UKF based approach to enhance the robot pose $\mathbf{x} = \left[x_k\ y_k\ \theta_k\right]^T$ estimate, the inputs of the localization system are considered the wheels angular velocities $\mathbf{u} = \left[\omega_{Rk}\ \omega_{Lk}\right]^T$ as a relative measurement system. The map, is represented by four landmarks, $z_k = \left[m_x^{[i]}\ m_y^{[i]}\right]^T$, $i = 1$ to 4, as an absolute measurement system obtained from a detected key point and matched with a database stored images. The UKF-Localization algorithm includes three main stages as seen in Fig. 4.

Δ represents a one sample time delay of the state estimate obtained from the UKF algorithm.

First, a set of sigma points are voted to express the probability distribution of the random variable, the nonlinear function is then applied to each sigma point and form the joint distribution function of the motion model as follows (5):

$$\begin{pmatrix} \chi_{k-1} \\ \chi_k \end{pmatrix} \sim N\left(\begin{pmatrix} \chi_{k-1} \\ \hat{\chi}_k^- \end{pmatrix}, \begin{pmatrix} \mathbf{P}_{k-1} & \mathbf{0} \\ \mathbf{0} & \mathbf{P}_{\chi\chi}^- \end{pmatrix}\right) \tag{5}$$

where $\mathbf{P}_{k-1}$ is the initial covariance matrix and $\mathbf{P}_{\chi\chi}^-$ is the priori covariance matrix of the motion model [21].

Next, propagate the new points through the measurement model and calculate the predicted mean $\hat{\mathbf{z}}_k$, the predicted covariance of the measurement $\mathbf{P}_{zz}$ and the cross-covariance of the motion and measurement $\mathbf{P}_{\chi z}$ as given by (6):

$$\begin{pmatrix} \chi_k \\ \mathbf{z}_k \end{pmatrix} \sim N\left(\begin{pmatrix} \hat{\chi}_k^- \\ \hat{\mathbf{z}}_k \end{pmatrix}, \begin{pmatrix} \mathbf{P}_{\chi\chi}^- & \mathbf{P}_{\chi z} \\ \mathbf{P}_{z\chi} & \mathbf{P}_{zz} \end{pmatrix} \right) \tag{6}$$

The last stage is the Correction Update; the 4WDDMR corrects its pose by fusing the priori belief comes from the relative measurement system, which is onboard sensors with the probability of making exact observation using absolute measurement system, which comes from a detected landmark [15].

Table 1, shows the pseudo code for UKF-localization algorithm [24].

4 Simulation Results and Discussion

This section provides a comparison study and simulation results of two probabilistic based approaches namely; EKF and UKF. The two algorithms are implemented using C language and augmented with MATLAB/SIMULINK model of a 4WDDMR [17].

With reference to Fig. 1 it is assumed that the robot has to follow a square path, in which the initial robot pose robot is predefined and known. Four landmarks are also determined in the robot environment. As seen in Fig. 5, the mobile robot tracks correctly the desired path based on UKF approach, but there seem to be a big divergence for the case of EKF. Figure 6 shows the trajectory tracking error for the UKF based approach. As seen the peak error in Cartesian x and y coordinates is almost less than 0.1 m, which is due to the system dynamic response as the robot react to the desired pose command. It is noted that the UKF based approach provides highest DoB in which the robot is certain to localize correctly.

The desired path plan is given as (7):

$$\begin{bmatrix} x_d^{[i]} \\ y_d^{[i]} \\ \theta_d^{[i]} \end{bmatrix} = \begin{bmatrix} 0 & 2 & 2 & 0 & 0 \\ 0 & 0 & 2 & 2 & 0 \\ 0 & 0 & 90 & -90 & 0 \end{bmatrix}_{i=1,\ldots,5} \tag{7}$$

Based on the planned trajectory given in (7), the robot executes the next desired pose command as long as it locates correctly on the desired pose. When the robot considers its degree of belief (DoB) is high and certain it goes ahead to execute the next desired pose in the path plan.

Figure 7 shows the error in the trajectory tracking when using EKF based approach. As noted, the error is unaffordable since the robot is lost and not capable to localize correctly, since the DoB based on EKF approach is not fully trusted, with

Table 1 Pseudo Code for UKF-localization algorithm

Unscented Kalman filter $(\mathbf{x}_{k-1}, \mathbf{P}_{k-1}, \mathbf{u}_k, \mathbf{z}_k)$

Initialization:

$$\chi_0 = \begin{bmatrix} \mathbf{x}_0 & 0 & 0 \end{bmatrix}^T$$

$$\mathbf{p}_{k-1} = \begin{bmatrix} \mathbf{P}_0 & \mathbf{0} & \mathbf{0} \\ \mathbf{0} & \mathbf{Q} & \mathbf{0} \\ \mathbf{0} & \mathbf{0} & \mathbf{R} \end{bmatrix}$$

For each sample time k do:

Prediction:

1- Compute 2n + 1 Sigma Points, where γ is a scalar parameter:

$$\chi_{k-1} = \begin{bmatrix} \mathbf{x}_{k-1} & \mathbf{x}_{k-1} + \gamma\sqrt{\mathbf{p}_{k-1}} & \mathbf{x}_{k-1} - \gamma\sqrt{\mathbf{p}_{k-1}} \end{bmatrix}^T$$

2- Apply The Kinematic Model:

$$\chi_k = f(\chi_{k-1}, \mathbf{u}_k, 0)$$

3- Compute Sigma Points Mean and Covariance:

$$\hat{\chi}_k^- = \sum_{i=0}^{2n} \omega_m^{[i]} \chi_k^{[i]}$$

$$\mathbf{P}_{\chi\chi}^- = \sum_{i=0}^{2n} \omega_c^{[i]} \left(\chi_k^{[i]} - \hat{\chi}_k^-\right)\left(\chi_k^{[i]} - \hat{\chi}_k^-\right)^T$$

Where $\omega_m^{[i]}$, $\omega_c^{[i]}$ define by the algorithm.

4- Apply The Measurement Model:

$$\mathbf{z}_k = h(\chi_k^{[i]}, \mathbf{m}_k, 0)$$

5- Compute the New Sigma Points Mean and Covariance:

$$\hat{\mathbf{z}}_k = \sum_{i=0}^{2n} \omega_m^{[i]} \mathbf{z}_k^{[i]}$$

$$\mathbf{P}_{zz} = \sum_{i=0}^{2n} \omega_c^{[i]} \left(\mathbf{z}_k^{[i]} - \hat{\mathbf{z}}_k\right)\left(\mathbf{z}_k^{[i]} - \hat{\mathbf{z}}_k\right)^T$$

$$\mathbf{P}_{\chi z} = \sum_{i=0}^{2n} \omega_c^{[i]} \left(\chi_k^{[i]} - \hat{\chi}_k^-\right)\left(\mathbf{z}_k^{[i]} - \hat{\mathbf{z}}_k\right)^T$$

Correction: The UKF gain $\mathbf{K}_k$ is chosen to minimize the errors and calculate the posterior mean and covariance $(\hat{\chi}_k, \mathbf{P}_{\chi\chi})$

$$\mathbf{K}_k = \mathbf{P}_{xz}\mathbf{P}_{zz}^{-1}$$

$$\hat{\chi}_k = \hat{\chi}_k^- + \mathbf{K}_k\left(\mathbf{z}_k - \hat{\mathbf{z}}_k\right)$$

$$\mathbf{P}_{xx} = \mathbf{P}_{xx}^- - \mathbf{K}_k\mathbf{P}_{zz}\mathbf{K}_k^T$$

return $(\hat{\chi}_k, \mathbf{P}_{xx})$

end function

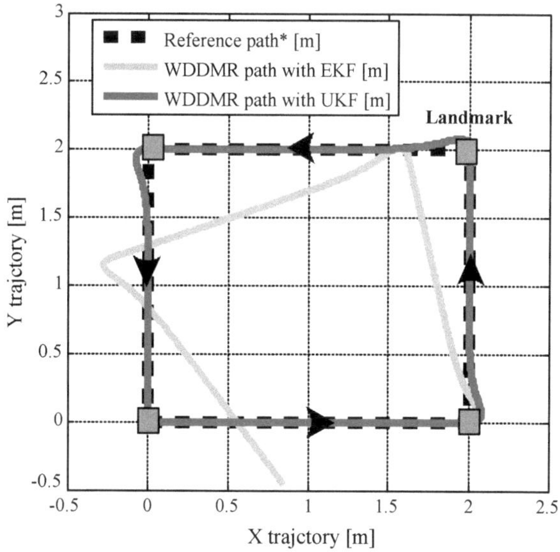

Fig. 5 Square Path EKF/UKF-localization performance

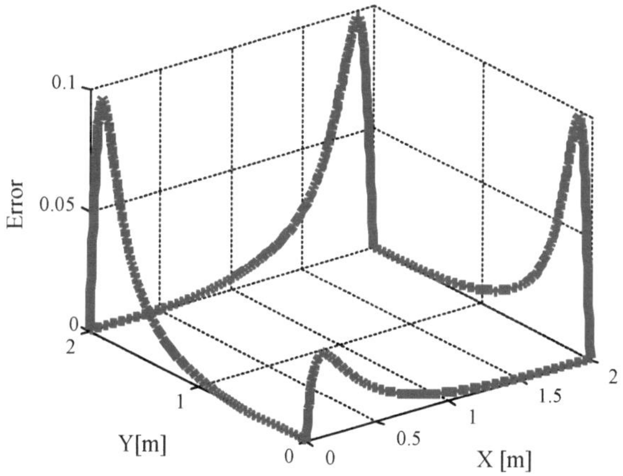

Fig. 6 Square Path UKF-localization error

reference to Fig. 5 blue trajectory, it is seen that the robot is even not able to return to the origin pose $\begin{bmatrix} 0 & 0 & 0 \end{bmatrix}^T$, and diverged to an uncertain pose.

Figure 8 shows the robot trajectory tracking in which the desired path is a circular with a diameter of a 2 m. It is notable that initially, the robot is located in the coordinate $\begin{bmatrix} x \\ y \\ \theta \end{bmatrix} = \begin{bmatrix} 0 \\ 0 \\ 0 \end{bmatrix}$, but the desired trajectory begins at, $\begin{bmatrix} x_d^{[i]} \\ y_d^{[i]} \\ \theta_d^{[i]} \end{bmatrix} = \begin{bmatrix} 0 \\ 2 \\ 180 \end{bmatrix}$. As seen the UKF based approach is smoothly tracked the desired trajectory blue curve. For the

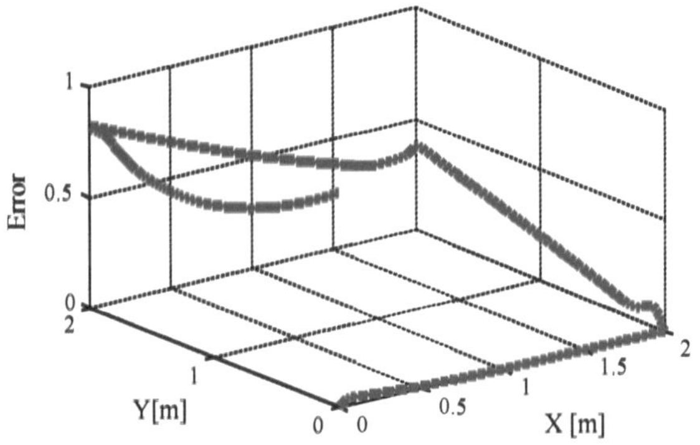

Fig. 7 Square Path EKF-localization error

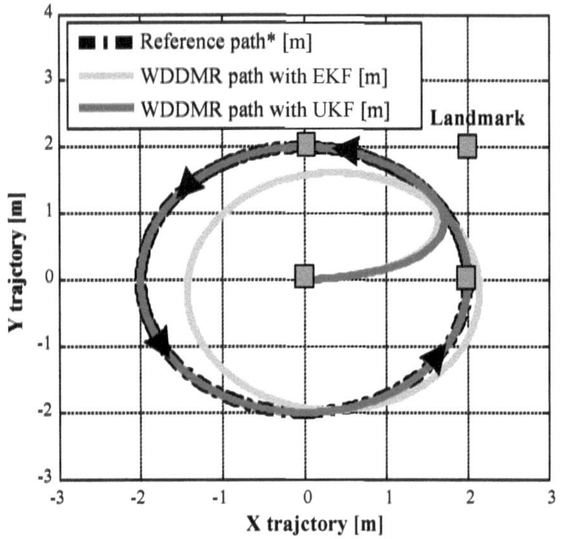

Fig. 8 Circular Path EKF/UKF-localization performance

EKF approach there was a lack of high DoB and the robot localization was lost and diverged from the desired trajectory.

Figure 9 shows the trajectory tracking error for the case of UKF approach. As seen, the error is largest at the start of the robot motion due to different initialization of both the desired robot pose and the actual one.

But the error converged to zero as the robot detected a landmark for which the DoB was highest and almost the trajectory tracking error was around zero.

Figure 10 shows the trajectory tracking error of a circular trajectory tracking with EKF approach. As seen, starting with similar tracking error as the case for the UKF approach see Fig. 9, but with no chance of correction due to uncertain DoB of the robot localization. As noted in Fig. 10 the tracking error was highest and no way to converge to zero. As seen, the robot moved away from the desired position.

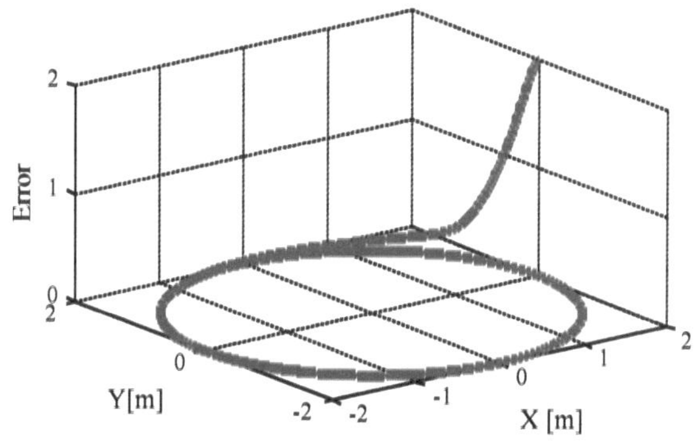

Fig. 9 Circular Path UKF-localization error

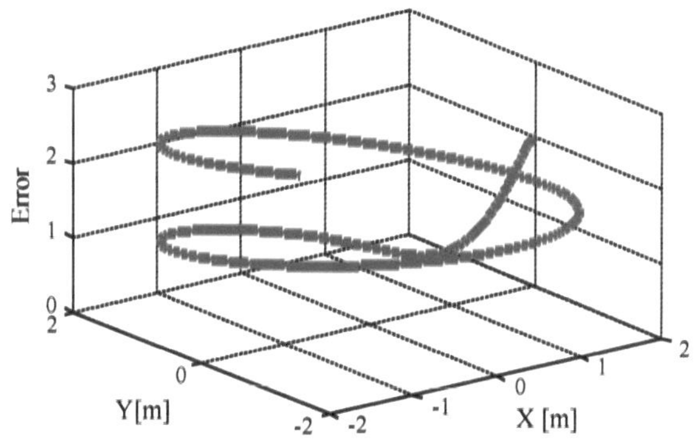

Fig. 10 Circular Path EKF-localization error

5 Conclusions

In this paper, a comparison study between EKF and UKF assisted SLAM for 4WDDMR localization using predifined landmarks has been discussed. Simulation results obtained from the developed 4WDDMR model built in MATLAB/SIMULINK, with UKF algorithm written in C-code have demonstrated very good tracking performance of the desired trajectory tracking. Square and circular trajectory tracking have been examined using both EKF and UKF assisted localization approaches. The problem of system linearization used with EKF approach has led to bias robot pose estimation, and ultimately resulted in inaccurate robot localization. The degree of belief (DoB) obtained from EKF approach has not been reliable and not fully trusted, for which the robot was not capable to localize correctly around its desired pose. However, enhancement of the robot pose estimation assisted with the predefined landmarks and UKF has been obtained. This has guaranteed high and trusted DoB in which the robot trajectory tracking error has been shown converging to zero. It is worth mentioning that the UKF based approach has exhibited a superior performance localization and tracking compared to EKF approach.

References

1. Beinhofer M, Müller J, Krause A, Burgard W (2013) Robust landmark selection for mobile robot navigation. In: Intelligent robots and systems (IROS), 2013 IEEE/RSJ international conference, (Nov 2013)
2. Se S, Lowe D, Little J (2002) Mobile robot localization and mapping with uncertainty using scale—invariant visual landmarks. Int J Robotics Res 21(8):735–758
3. Park S, Jung S, Song Y, Kim H (2008) Mobile robot localization in indoor environement using scale—invariant visual landmarks. In: 18th IAPR international conference on pattern recognition
4. Owen C, Nehmzow U (1998) Landmark—based navigation for a mobile robot. MIT Press
5. Merwe R (2004) Sigma-point Kalman filters for probabilistic inference in dynamic state-space models. PhD thesis
6. Liu Yong (2010) Navigation and control of mobile robot using sensor fusion. Robot Vision, InTech
7. Jensfelt P (2001). Approaches to mobile robt localization in indoor environement. Doctoral thesis, Royal Institute of Technology (KTH)
8. Park J, Teler S (2014) Motion compatibility for indoor localization. In: Computer sience and artifitial intelligence laboratory, massachusette institute of technology, Cambridge
9. Kerl C, Sturm J, Cremers D (2014) Dense visual SLAM for RGB-D Camera. IEEE
10. Park J, Lee S (2009) Correction robot pose for SLAM based on Extended Kalman filter in rough surface environement. Int J Adv Robotic Syst 6(2):67–72
11. Klančar G, Teslić L, Škrjanc I (2013) Mobile-robot pose estimation and environment mapping using an extended Kalman filter. Int J Syst Sci
12. M. Rhudy, Gu Y, Napolitano MR (2013) An analytical approach for comparing linearization methods in EKF and UKF. Int J Adv Robotic Syst, InTech
13. Negenborn R (2003) Robot localization and Kalman filters on finding your position in a noisy world. MSc thesis. (Utrecht University)
14. Corke P (2001) Robotics vision and control fundamental algorithms in MATLAB. Springer Tracts in Advanced Robotics, vol 73
15. Thrun S, Fox D, Burgard W (2006) Probabilistic Robotics. Massachusttes Institute of Technology
16. Siegwart R, Nourbakhsh IR (2004) Introduction to autonomous mobile robots. The MIT Press, Cambridge, Massachusetts London
17. Kass Hanna D, Joukhadar A (2015) A novel control-navigation system-based adaptive optimal controller & EKF localization of DDMR. Int J Adv Res Artif Intell (IJARAI) 4(5):29–37
18. Hayet JB, Lerasle F, Devy M (2002) A visual landmark framwork for indoor mobile robot navigation. In: International conference on robotics & automation, Washington, DC
19. Long AW, Wolfe KC, Mashner MJ, Chirikjian GS (2013) The banana distribution is Gaussian: A localization study with exponential coordinates. Robotics: Science and Systems VIII; MIT Press, Cambridge, MA, USA, pp 265–272
20. Houshangi N, Azizi F (2005) Accurate mobile robot position determination using unscented Kalman filter. IEEE, Saskatoon
21. Särkkä S (2011) Bayesian filtering and smoothing. Cambridge University Press, vol 3
22. Kandepu R, Foss B, Imsland L (2007) Applying the unscented Kalman filter for nonlinear state estimation. Process Control
23. Haykin S (2001) Kalman filtering and neural networks. A Wiley—Interscience Publication
24. Press WH, Teukolsky SA, Vetterling WT, Flannery BP (2007) Numerical recipes the art of scientific computing. Cambridge University Press, UK

Trajectory Tracking Control of a Quadcopter UAV Using Nonlinear Control

Abdulkader Joukhadar, Mohammad AlChehabi, Christoph Stöger and Andreas Müller

Abstract Recently trajectory tracking control of a quadcopter has been paid attention by academic and industry. This paper proposes two different strategies for trajectory tracking control of a quadcopter system implementing nonlinear control theory. The first approach is based on the integral backstepping technique, the second proposed one is an LQI (Linear Quadratic Integral) optimal controller with a feedback linearization so as to deal with the nonlinearity and the coupling components of the quadcopter state variables. The control laws for trajectory tracking using the proposed two strategies were validated by simulation and experimental results obtained from a quadcopter test bench. Simulation results show a comparison between the performance of each of the two control laws depending on the nonlinear model of the quadcopter system under investigation; the trajectory tracking has been achieved properly for different types of trajectories in presence of unknown disturbances. Simulation and practical results have shown coincided tracking with the command signals of the desired attitude. Superior tracking control has been exhibited with the proposed LQI optimal controller. It has been also noted that the proposed control approach exhibits an inherited decoupling control action, for which the control of one axis angle has relieved the dynamic coupling effect on the other two axes. Furthermore, intensive practical results have demonstrated the robustness of the proposed controller.

1st ICAMMRMS, Beirut, Lebanon, 17–19 October, 2017.

A. Joukhadar (✉) · M. AlChehabi
University of Aleppo, Aleppo, Syria
e-mail: ajoukhadar@alepuniv.edu.sy

M. AlChehabi
e-mail: mohammad.alchehabi@hotmail.com

C. Stöger · A. Müller
Johannes Kepler University, Linz, Austria
e-mail: christoph.stoeger@jku.at

A. Müller
e-mail: a.mueller@jku.at

R. Rizk and M. Awad (eds.), *Mechanism, Machine, Robotics and Mechatronics Sciences*, Mechanisms and Machine Science 58,
https://doi.org/10.1007/978-3-319-89911-4_20

Keywords Backstepping · Feedback linearization · Quadcopter · LQI

1 Introduction

In last decade, academic and industry have shown interest in Unmanned Aerial Vehicles (UAVs) Systems, due to its wide applications in many civil domain and military such as photographing and reconnaissance [1–4]. The quadcopter, a four-rotors helicopter with (+) or (x) shapes is one of these UAVs types and also classified as a flying robot. It has many advantages over a single rotor conventional helicopter. Firstly, it has the ability of Vertical Take-off and Landing (VTOL), hovering and achieving acrobatic manoeuvres. Secondly, decreasing in power consumption, since it consists of small-scale 4-motor instead of one main large-scale motor for thrusting and another small-scale motor in tail of the helicopter for pitching. The motion of the quadcopter depends on the magnitude of the forces and torques produced by the propellers [1, 2].

Integral backstepping is one of the most widely used nonlinear control techniques, this is due to its advantages including: being a recursive approach which depends on a Lyapunov Function for deriving a desired control law; flexibility to avoid cancellation of useful nonlinearities and pursuing the objective of stabilization and tracking, rather than that of linearization [5, 6]. On the other hand, input/output linearization method is also one of the major nonlinear design tools, which is used for decoupling control of the coupling components among state variables of a nonlinear system, in addition it allows designing and incorporating a linear controller for a nonlinear system with the feedback linearization laws [5]. The LQI optimal controller is considered one of the most widely implemented optimal controllers used with linear or linearized systems due to its ability for achieving stability and tracking, and eliminating steady state static errors [7].

Recently, several control methods have been tested on quadcopter control systems for achieving high stability and tracking control performance. In [8] an integral backstepping method for full position control is applied. A nonlinear H_∞ controller is implemented to control the attitude and a backstepping controller for controlling the position of the quadcopter is implemented. However, Hamilton-Jacobi-Bellman-Isaacs partial derivative equations (HJBI PDEs) are used instead of Riccati equations for the case of linear H_∞ control. The main problem in the nonlinear case of the H_∞ is the absence of a general method to solve the HJBI PDEs [9]. In [10] a modified backstepping approach is proposed to control the attitude and position of the quadcopter where the main contribution is the reduce of the number of the controller tuning parameters in the control laws. In [11] a nonlinear control design is combined with an on-line parameter adaptation to develop the control law in the presence of parameter uncertainties. The dynamic performance is compared with the performance of the sliding mode controller. In [12] Fuzzy backstepping Sliding Mode Controllers are designed for the quadcopter. However, this control technique is similar to the conventional integral backstepping technique where the error signals are determined as the sliding manifold; moreover, a fuzzy controller is added instead of the sign function in the control law. In [13] adaptive backstepping controllers are designed and tested experimentally.

In [14] LQR controllers for the attitude and position of a quadcopter are tested experimentally. In [15] two full linear controllers for the attitude and position of a quadcopter are implemented using simplified nonlinear model. The first one is an LQI controller and the second is an H_∞ controller. In [16] a linear quadratic tracking (LQT) controller is derived depending on the discrete linear model of a quadcopter, where the difference between LQT and LQR is the gain values imposed in the LQT controller, which are online tuned. In [17] a robust optimal controller depending on L_l-norm for tracking purposes is proposed, in which the simulation results are tested practically.

On the other hand, academic and industry have concentrated on tracking purposes for the quadcopter. In [18] LQ controllers with feedback linearization are applied for the attitude and altitude of the quadcopter and backstepping controllers of x & y position control are used. [19] shows nonlinear control laws applied for optimal trajectory tracking, control laws link between the system outputs and its inputs are derived depending on differential flatness method. Where [20] presents the trajectory tracking control of a quadcopter which is achieved relying on integral backstepping method. In [21] backstepping and sliding mode control methods are used for deriving control laws of a quadcopter for trajectory tracking purposes. The remaining sections of this paper are organized as follows; Sect. 2 discusses the Dynamic Modelling of the Quadcopter. Deriving the control laws has been shown in Sect. 3. Section 4, provides simulation and experimental results. Conclusion and future work are discussed in Sect. 5.

2 Modelling of the Quadcopter

The dynamic model of the quadcopter is mostly derived either using Newton's laws or the Lagrange formula [3–5]. This paper considers the Newton-Euler equations for the dynamic model derivation, which are given as follows:

$$\vec{F} = m\dot{V} + \omega \times m\vec{V} \tag{1}$$

$$\vec{T} = I\dot{\omega} + \omega \times I\vec{\omega} \tag{2}$$

The state variables of the quadcopter are defined as, $X = [x, y, z, \dot{x}, \dot{y}, \dot{z}, \phi, \theta, \psi, \dot{\phi}, \dot{\theta}, \dot{\psi}]^T$ where, $\zeta = [x, y, z]^T$ is the position described in the inertial coordinate frame B, $V = [\dot{x}, \dot{y}, \dot{z}]^T$ is the translational velocity, $\eta = [\phi, \theta, \psi]^T$ are the roll-pitch-yaw angles describing the attitude of the quadcopter, $\dot{\eta} = [\dot{\phi}, \dot{\theta}, \dot{\psi}]^T$ are the Euler angle rates and $\omega = [p, q, r]^T$ the angular velocity of the quadcopter described in the body fixed frame A.

With $\omega = R_r \dot{\eta}$ and $R_r = \begin{bmatrix} 1 & 0 & -s\phi \\ 0 & c\phi & c\theta.s\phi \\ 0 & -s\phi & c\theta.c\phi \end{bmatrix}$.

For small Euler angles, the angular velocity of the quadcopter is considered as $\omega = \dot{\eta}$ [2, 5]. Equation (1) can be formulated as given in (3), considering the term $\omega \times m\vec{V} = 0$ since the velocity of the quadcopter body is described in an inertial frame thus:

$$m\dot{V} = -mG + {}^{B}R_{A}F \tag{3}$$

where: ${}^{B}R_{A}$ is the transformation matrix.

$${}^{B}R_{A} = \begin{bmatrix} c\psi c\theta & s\phi.s\theta.c\psi - c\phi s\psi & c\phi.s\theta.c\psi + s\phi.s\psi \\ s\psi c\theta & s\phi.s\theta.s\psi + c\phi c\psi & c\phi.s\theta.s\psi - s\phi.c\psi \\ -s\theta & s\phi c\theta & c\phi c\theta \end{bmatrix} \tag{4}$$

m: the mass of the quadcopter in Kg. $F = [0, 0, U_1]^T$ is the magnitude of the thrust force, $G = [0, 0, g]^{\mathrm{T}}$ the gravity rate g in m.s^{-2}. With,

$$U_1 = f_1 + f_2 + f_3 + f_4 \tag{5}$$

where $f_i = b\omega_i^2$ is the thrust force produced by propeller i with thrust coefficient b in N.s^2. ω_i is the angular speed of motor i. Writing the Eq. (3) in matrix form, one gets:

$$m \begin{bmatrix} \ddot{x} \\ \ddot{y} \\ \ddot{z} \end{bmatrix} = -m \begin{bmatrix} 0 \\ 0 \\ g \end{bmatrix} + {}^{B}R_{A} \begin{bmatrix} 0 \\ 0 \\ U_1 \end{bmatrix} \tag{6}$$

On the other hand, the equation of the angular motion (2) is formulated as follows:

$$I\dot{\omega} = \tau + \tau_g - \omega \times I\vec{\omega} \tag{7}$$

$$\tau = [\tau_\phi, \tau_\theta, \tau_\psi]^T = [U_2, U_3, U_4]^T$$

$$\tau_g = \left[J_r \Omega_r \dot{\theta} \; -J_r \Omega_r \dot{\phi} \; 0 \right]^T \tag{8}$$

$$\tau_\phi = l(f_4 - f_2) \tag{9}$$

$$\tau_\theta = l(f_3 - f_1) \tag{10}$$

$$\tau_\psi = T_1 - T_2 + T_3 - T_4 \tag{11}$$

$T_i = d\omega_i^2$ is the drag torque produced by propeller i in N.m with corresponding drag coefficient $\boldsymbol{d}$ in Nm^2.s. τ_g represents the gyroscopic effect in the term of orientation change in the propellers plane. $\boldsymbol{l}$ is the distance between center of the quadcopter and center of propeller in m. $\omega \times I\vec{\omega}$ is the gyroscopic effect, I is the inertia matrix, where I_{xx}, I_{yy}, I_{zz} are moments of inertia about x, y, z axes, respectively in Kg.m^2.

$$\begin{bmatrix} I_{xx} & 0 & 0 \\ 0 & I_{yy} & 0 \\ 0 & 0 & I_{zz} \end{bmatrix} \begin{bmatrix} \ddot{\phi} \\ \ddot{\theta} \\ \ddot{\psi} \end{bmatrix} = \begin{bmatrix} U_2 \\ U_3 \\ U_4 \end{bmatrix} + \begin{bmatrix} J_r \Omega_r \dot{\theta} \\ -J_r \Omega_r \dot{\phi} \\ 0 \end{bmatrix} - \begin{bmatrix} \dot{\phi} \\ \dot{\theta} \\ \dot{\psi} \end{bmatrix} \times \begin{bmatrix} I_{xx} & 0 & 0 \\ 0 & I_{yy} & 0 \\ 0 & 0 & I_{zz} \end{bmatrix} \begin{bmatrix} \dot{\phi} \\ \dot{\theta} \\ \dot{\psi} \end{bmatrix} \tag{12}$$

where J_r is the inertia of propeller and Ω_r is the sum of the four motors angular speed. Based on the above derivation and discussion, the equations of motion can be written as follows (13):

$$\begin{aligned} \ddot{x} &= (c\phi s\theta c\psi + s\phi s\psi)\frac{U_1}{m} \\ \ddot{y} &= (c\phi s\theta s\psi - s\phi s\psi)\frac{U_1}{m} \\ \ddot{z} &= (c\phi c\theta)\frac{U_1}{m} - g \\ \ddot{\phi} &= \frac{I_{zz} - I_{yy}}{I_{xx}}\dot{\theta}\dot{\psi} + \frac{J_r}{I_{xx}}\Omega_r\dot{\theta} + \frac{1}{I_{xx}}U_2 \\ \ddot{\theta} &= \frac{I_{zz} - I_{xx}}{I_{yy}}\dot{\phi}\dot{\psi} - \frac{J_r}{I_{yy}}\Omega_r\dot{\phi} + \frac{1}{I_{yy}}U_3 \\ \ddot{\psi} &= \frac{I_{xx} - I_{yy}}{I_{zz}}\dot{\phi}\dot{\theta} + \frac{1}{I_{zz}}U_4 \end{aligned} \tag{13}$$

$$U = \begin{bmatrix} U_1 \\ U_2 \\ U_3 \\ U_4 \end{bmatrix} = \begin{bmatrix} b & b & b & b \\ 0 & -bl & 0 & bl \\ -bl & 0 & bl & 0 \\ d & -d & d & -d \end{bmatrix} \begin{bmatrix} \omega_1^2 \\ \omega_2^2 \\ \omega_3^2 \\ \omega_4^2 \end{bmatrix} \tag{14}$$

3 Quadcopter Control

This paper presents two nonlinear control methods derivations in order to control the quadcopter and achieve a high dynamic performance for trajectory tracking. The first method is based on the integral backstepping technique, and the second one relies on feedback linearization with an LQI optimal control.

3.1 *Integral Backstepping*

Control laws of the attitude and position of the quadcopter are derived using the integral backstepping approach. This approach depends on determining the error between the desired input and the system actual output, then outlining a Lyapunov function and determining virtual controls to make the derivative of the proposed Lyapunov function negative definite [6].

3.1.1 Altitude Control

The control laws for the altitude of the quadcopter is derived as follows. The altitude error:

$$e_1 = z_d - z \tag{15}$$

and the Lyapunov function is chosen as

$$V_1 = \frac{1}{2}e_1^2 \tag{16}$$

with $\dot{V}_1 = e_1\dot{e}_1$, where

$$\dot{e}_1 = \dot{z}_d - \dot{z} = \dot{z}_d - v_z \tag{17}$$

If the term $k_I e_I$ is added and subtracted to the $\dot{V}_1$ function, where $k_I{>}0$, then this yields

$$\dot{V}_1 = e_1\dot{e}_1 = e_1(\dot{z}_d - V_z + k_1e_1 - k_1e_1) \tag{18}$$

$$\dot{V}_1 = -k_1e_1^2 + e_1(\dot{z}_d - V_z + k_1e_1) \tag{19}$$

The term $\dot{z}_d - v_z + k_1e_1$ must vanish for a negative definite derivative of the Lyapunov function. That can be achieved by choosing the virtual control v_z such that

$$v_{z_d} = \dot{z}_d + k_1e_1 + c_1\int e_1dt \tag{20}$$

Similar steps are repeated here to derive the control law,

$$e_2 = v_{z_d} - v_z \tag{21}$$

The proposed Lyapunov function is

$$V_2 = \frac{1}{2}e_2^2 \tag{22}$$

The 1st order derivative of the proposed function is

$$\dot{V}_2 = e_2 \dot{e}_2 \tag{23}$$

Using a similar strategy as for v_{z_d} results

$$\dot{V}_2 = -k_2 e_2^2 + e_2(\dot{v}_{z_d} - \frac{c\varphi c\theta}{m} U_1 + g + k_2 e_2) \tag{24}$$

$$U_1 = \frac{m}{c\phi c\theta}\{\ddot{z}_d + k_1 \dot{e}_1 + g + k_2 e_2\} \tag{25}$$

3.1.2 Attitude Control

The control laws of the attitude of the quadcopter were derived in this section depending on integral backstepping method as follows:

$$U_2 = \frac{1}{b_1}\{\ddot{\phi}_d + k_3 \dot{e}_3 - a_1 \dot{\theta}\dot{\psi} - a_2 \dot{\theta}\Omega_r + k_4 e_4\} \tag{26}$$

$$U_3 = \frac{1}{b_2}\{\ddot{\theta}_d + k_5 \dot{e}_5 - a_2 \dot{\theta}\dot{\psi} + a_4 \dot{\theta}\Omega_r\} \tag{27}$$

$$U_4 = \frac{1}{b_3}\{\ddot{\psi}_d + k_7 \dot{e}_7 - a_5 \dot{\phi}\dot{\theta}\} \tag{28}$$

3.1.3 Position Control

Since the motion of the quadcopter on the x axis depends on θ angle, whereas the motion on the y axis depends on ϕ angle, thus ϕ and θ angles have been considered as the outputs of x, y control laws, it is remarkable to mention that small Euler angles have not been considered in order to obtain the position control laws in this paper, which is significant to achieve high dynamic performance tracking control. The main contribution of this paper is deriving the position control laws from the quadcopter's model directly just by applying the procedure of the control approaches. Recalling Eqs. (13), for clarification, as follows

$$\ddot{x} = \frac{c\phi s\theta c\psi + s\phi s\psi}{m} U_1; \ \ddot{y} = \frac{c\phi s\theta s\psi - s\phi s\psi}{m} U_1; \tag{29}$$

At first, the integral backstepping method is applied to derive the control law for the motion of the quadcopter on the x axis, and then the following control law is obtained:

$$\theta_d = \arcsin\left(\frac{m}{c\phi.c\theta.U_1}\left\{\dot{v}_{x_d} - \frac{s\phi.s\psi}{m} U_1 + k_{10} e_{10}\right\}\right) \tag{30}$$

The control law for the motion of the quadcopter on the y axis is derived in a similar way, and then the following law is obtained:

$$\phi_d = -\arcsin\left(\frac{m}{c\psi.U_1}\left\{\dot{v}_{y_d} - \frac{c\phi.s\theta.s\psi}{m}U_1 + k_{12}e_{12}\right\}\right) \tag{31}$$

3.2 *Input/Output Linearization with LQI*

The feedback linearization method is used in this paper in order to decouple the state variables of the quadcopter. This will enable us to derive the LQ based control laws for the attitude, altitude and position of the quadcopter.

3.2.1 Altitude Control

The feedback linearization of the attitude is given as follows

$$Y_3 = z \tag{32}$$

$$\ddot{Y}_3 = \ddot{z} = \frac{c\phi.c\theta}{m}U_1 - g \tag{33}$$

$$U_1 = \frac{m}{c\phi.c\theta}(V_1 + g) \tag{34}$$

V_1 is a virtual input which is computed using LQI controller that will be presented in section C.

3.2.2 Attitude Control

The feedback linearization laws of the attitude are derived as follows

$$U_2 = \frac{I_{xx}}{l}\{-a_1\dot{\theta}\dot{\psi} - a_2\Omega_r\dot{\theta} + V_2\} \tag{35}$$

$$U_3 = \frac{I_{yy}}{l}\{-a_3\dot{\phi}\dot{\psi} + a_4\Omega_r\dot{\phi} + V_3\} \tag{36}$$

$$U_4 = \frac{I_{zz}}{l}\{-a_5\dot{\phi}\dot{\theta} + V_4\} \tag{37}$$

The previous control laws linearize the mapping between the derivatives of the flat outputs $Y_4 = \phi$, $Y_5 = \theta$, $Y_6 = \psi$ *and the virtual controls* V_2, V_3, V_4. The latter are again computed using an LQI optimal controller.

Where $a_1 = \frac{(I_{yy}-I_{zz})}{I_{xx}}$, $a_2 = \frac{J_r}{I_{xx}}$, $a_3 = \frac{(I_{zz}-I_{xx})}{I_{yy}}$, $a_4 = \frac{J_r}{I_{yy}}$, and $a_5 = \frac{(I_{zz}-I_{xx})}{I_{yy}}$.

3.2.3 Position Control

ϕ and θ angles are computed by the control laws of x and y motion, as it is done in the integral backstepping approach. The control laws are obtained as follows:

$$\theta_d = \arcsin\left(\frac{m}{c\phi.c\theta.U_1}\left\{\dot{v}_{x_d} - \frac{s\phi.s\psi}{m}U_1 + V_5\right\}\right) \tag{38}$$

$$\phi_d = -\arcsin\left(\frac{m}{c\psi.U_1}\left\{\dot{v}_{y_d} - \frac{c\phi.s\theta.s\psi}{m}U_1 + V_6\right\}\right) \tag{39}$$

V_5, V_6 are again LQI optimal controller inputs.

3.3 *LQI Optimal Control*

The main aim of the optimal control is to choose the control feedback such, that it minimizes a certain cost function. In the case of the linear quadratic regulator (LQR), the cost function is defined by [7]

$$J = \int_0^\infty (x^T Qx + u^T Ru)dt \tag{40}$$

Matrices Q and R represent costs on the state vector x and input vector u, respectively. LQRs are commonly applied to linear or linearized nonlinear systems, which are represented by state space models as follows

$$\begin{aligned} \dot{x} &= Ax + Bu \\ y &= Cx + Du \end{aligned} \tag{41}$$

The relation of the control law that minimizes the cost function J is given by

$$u = -K.x = -R^{-1}B^T P.x \tag{42}$$

With P, the covariance matrix represents the solution of the algebraic Riccati equation

$$A^T P + PA - PBR^{-1}B^T P + Q = 0 \tag{43}$$

The LQR controller provides good performance; however, it is not able to ensure fast tracking of dynamic references. In fact, there are many types of optimal followers [7], one of these is an LQI, with an integral action added, as shown in Fig. 1:

If the model of the linear system is extended by an error vector $\dot{z}$ such as

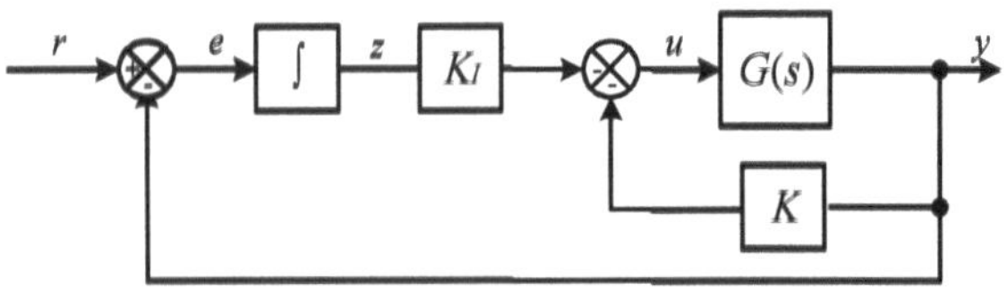

Fig. 1 LQI optimal controller structure

$$\dot{z} = r - y = r - (Cx + Du) \quad (44)$$

r refers to the command signal to be tracked, so the model can be written as

$$\begin{bmatrix} \dot{x} \\ \dot{z} \end{bmatrix} = \begin{bmatrix} A & 0 \\ -C & 0 \end{bmatrix} \begin{bmatrix} x \\ z \end{bmatrix} + \begin{bmatrix} B & 0 \\ D & I \end{bmatrix} \begin{bmatrix} u \\ r \end{bmatrix} \quad (45)$$

The control law $\bar{u}$ for the extended state space representation is given as follows

$$\bar{u} = -K\bar{x} - K_I\bar{z} \quad (46)$$

4 Implementation and Results

The proposed two approaches have been tested in MATLAB/Simulink based on the nonlinear quadcopter model of Eqs. (13), as well as experimental verification is also conducted. For Modelling and Simulation of the proposed approached, the simulation sample time was $T_s = 100$ μs, and Runge-Kutta fixed step integration has been utilized. Figures 3 and 4 exhibit the response of the trajectory tracking, where (a) labeled figures represent the system response using integral backstepping and (b) ones represent the system response using feedback linearization with LQI. Figure 2 demonstrates the desired trajectory and the actual trajectory tracking of the quadcopter considering a command heading angle $\psi = 0.4$ rad, and in presence of external disturbances imposed with the desired position commands in period between t = 15 s and t = 35 s. The initial point of the desired trajectory is (2, 0, 0), whereas the quadcopter initial position is (0, 0, 0). It is seen that the initial position of the quadcopter is a bit diverged from the initial position of the desired trajectory, where the actual trajectory of the quadcopter converged to the desired trajectory fast. Figure 3 shows the command signals and the responses for x, y and z-axes, the desired motion x and y motion were sinusoidal signals with 2 m magnitude and 0.5 rad.s^{-1} frequency and a ramp signal with 0.2 slope for the motion along the z axis.

The tracking errors of the motion of each axis are demonstrated in Fig. 4, it is noticed that the tracking error of motion on the x, y and z-axes converged to zero with a minor divergence between t = 15 s to t = 35 s, which is due to the presence of disturbance at this time instant. The attitude response of the quadcopter is shown in

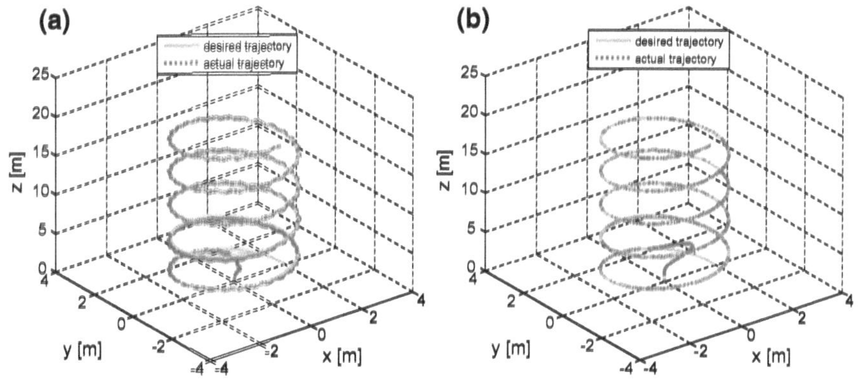

Fig. 2 Desired and actual trajectory. **a** Using integral backstepping. **b** Using feedback linearization with LQI

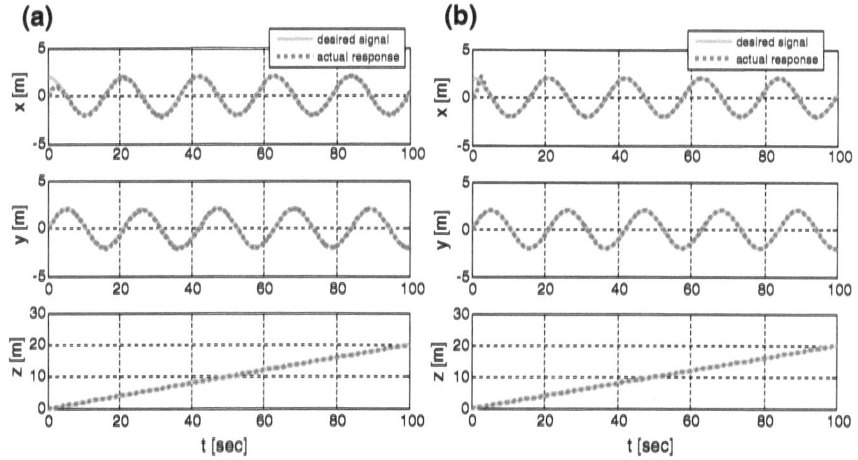

Fig. 3 Desired signals and actual response. **a** Using integral backstepping. **b** Using feedback linearization with LQI

Fig. 5, where the ϕ and θ responses represent the output of the motion on the x and y axes control laws, respectively.

The practical results of the attitude have been verified using Arduino MEGA board and the MPU6050 as an accelerometer and gyroscope. The components of the test bench assembled on a Flame Wheel F450 quadcopter. Figure 6. depicts the used practical control scheme to implement the control laws and acquire the data of the system response. The practical response of the pitch angle using integral backstepping technique has been demonstrated in Fig. 7. On the other hand, the practical response of the roll angle using both integral backstepping and feedback linearization with LQI has been depicted in Fig. 8a, b, respectively.

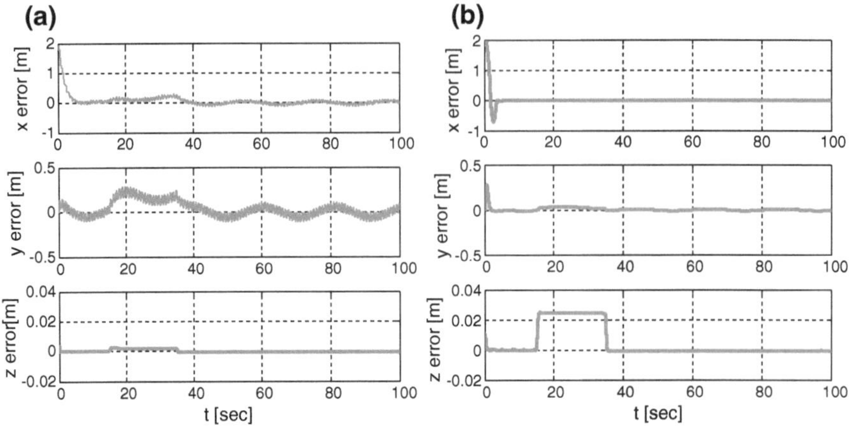

Fig. 4 Error signals. **a** Using integral backstepping. **b** Feedback linearization with LQI

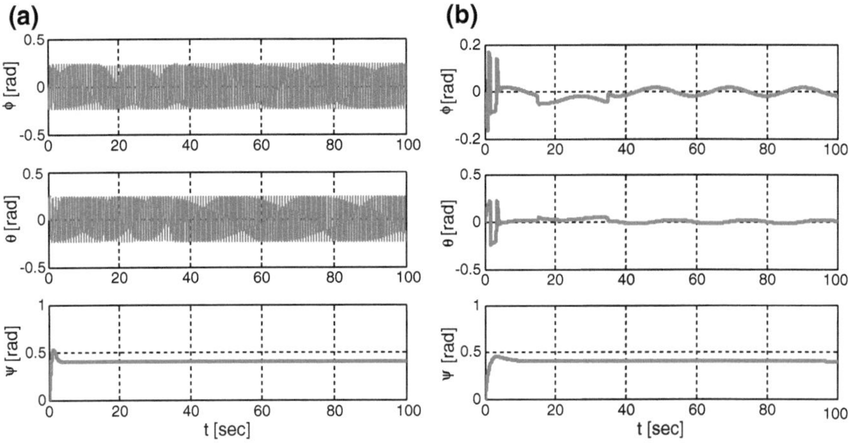

Fig. 5 Attitude response. **a** Using integral backstepping. **b** Feedback linearization with LQI

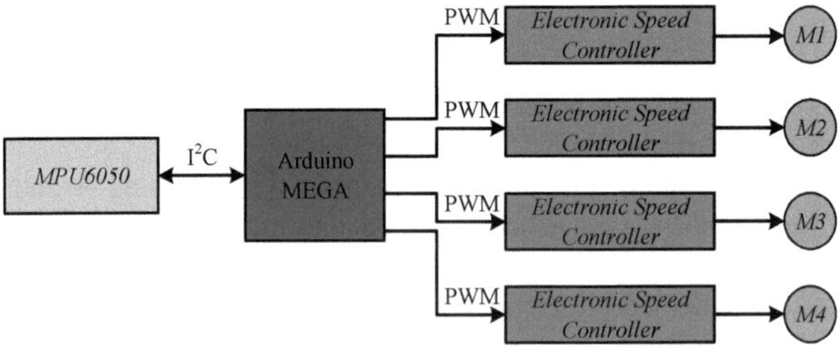

Fig. 6 Practical UAV control scheme

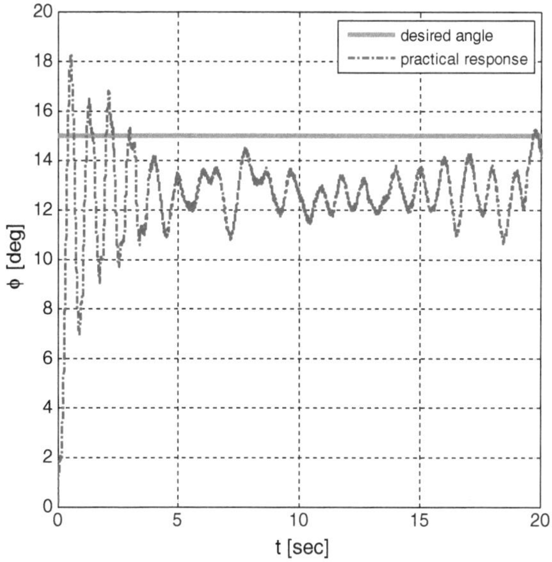

Fig. 7 Pitch practical response using integral backstepping

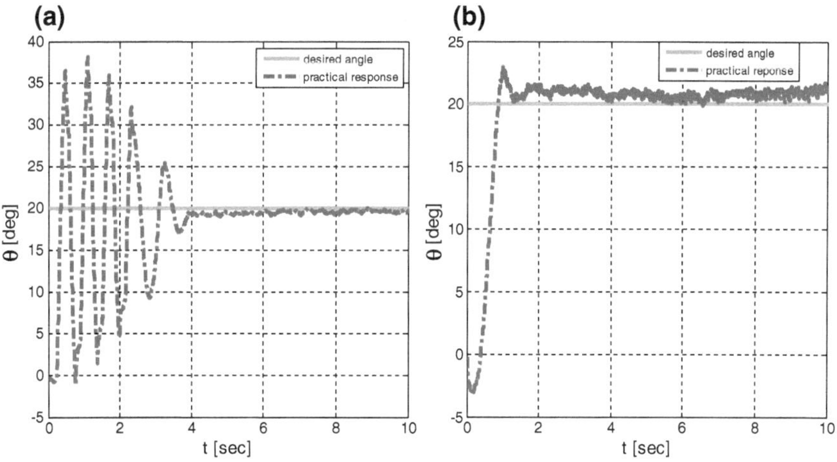

Fig. 8 Pitch practical response. (a) using integral backstepping. (b) feedback linearization with LQI

As seen in Fig. 8a, the pitch angle step response based on backstepping controller in the transient state is shown to be oscillating with undesired almost of ± 20 deg of overshoot and downshoot responses around the set point. However, the same angle response based on an LQI with feedback linearization approach is seen to be certain with normal exponential tracking response during the transient state.

5 Conclusions

The paper presented two nonlinear control approaches, which have been dealt with the task of trajectory tracking for a quadcopter UAV. At first, the nonlinear dynamic model of the quadcopter has been presented using Newton laws. The control scheme has been based on two control loops. An inner loop for attitude control and an outer loop for altitude and position control. Two different nonlinear control methods have been applied for both control loops. The first one was integral backstepping, and the second is the feedback linearization with an LQI optimal followers.

The dynamic performance of the proposed controllers of the tracking purpose was validated through simulation results in presence of external disturbances, whereas the control laws of the attitude was practically implemented using Arduino MEGA microcontroller board. Both methods exhibited excellent tracking and robustness to external disturbances, but the LQI with feedback linearization has shown superior dynamic performance in steady state response.

References

1. Nonami K, Kendoul F, Suzuki S et al. (2013) Autonomous flying robots. Springer, pp 1–75
2. Carillo L, López A, Lozano R et al. (2013) Quad rotorcraft control. Springer, pp 1–58
3. Castillo P, Lozano R, Dzul AE (2005) Modeling and control of mini flying machines. Springer, pp 1–56
4. Cui G, Chen B, Lee T (2011) Unmanned rotorcraft systems. Springer, pp 1–23
5. Khalil H (2002) Nonlinear systems. Prentice Hall, pp 505–603
6. Krstić M, Kanellakopoulos I, Kokotović P (1995) Nonlinear adaptive control design. Wiley, 1st edn, pp 19–86
7. Anderson B, Moore J (1989) Optimal control: linear quadratic method' Prentice-Hall, pp 1–99
8. Bouabdullah S (2007) Design and control of quadrotors with application to autonomous flying. PhD thesis No. 3727, EPFL
9. Raffo G, Ortega M, Rubio F (2008) Backstepping/nonlinear H_∞ control for path tracking of a quadrotor unmanned aerial vehicle' american control conference. Washington, USA, Seatle, pp 3356–3361
10. Saif A, Dhaifullah M, Al-Malki, M et al. (2012) Modified integral backstepping control of quadrotor' 9th international conference on systems. Signal and Devices
11. Diao C, Xian B, Gu X et al. (2012) Nonlinear control for an underactuated quadcopter unmanned aerial vehicle with parametric uncertainties' proceeding of 31st chinese conference on automatic control, Hefei, China, July 25–27, pp 998–1003
12. Khebbache H, Tadjine M (2013) Robust fuzzy backstepping sliding mode control for a quadcopter unmanned aerial vehicle. CEAI 15(2):3–11
13. Fang Z, Gao W (2012) adaptive backstepping control of an indoor micro-quadcopter. Res J Appl Sci Eng Technol 4(21):4216–4226
14. Santos O, Romero H, Salazar, S, et al. (2014) Discrete optimal control for a quadcopter UAV experimental approach. In: 2014 International conference on unmanned aircraft systems, Orlando, Fl, USA, May 27–30, 2014, pp 1138–1145
15. Araar O, Aouf N, Full linear control of a quadrotor UAV, LQ versus H_∞. UKACC International conference on control, loughborough, U.K, 9th–11th July 2014, pp 25–30
16. Suicmez E, Kutay A (2014) Optimal path tracking control of a quadrotor UAV' 2014 International conference on unmanned aircraft systems, Orlando, Fl, USA, May 27–30, pp 115–125

17. Satici A, Poonawla H, Spong M (2013) Robust optimal control of quadrotor UAVs' IEEE, vol 1, pp 79–93
18. Araar O, Aouf N (2014) Quadrotor control for trajectory tracking in presence of wind disturbances' UKACC International conference on control, loughborough, U.K, 9th–11th July 2014, pp 25–30
19. Mellinger D, Kumar V (2011) Minimum snap trajectory generation and control for quadcopters. IEEE International conference on robotics and automation, IEEE, May 2011, pp 2520–2525
20. Xu D, Guo L (2012) Modelling and trajectory tracking control of a quad-rotor UAV' the 2nd International conference on computer application and system modelling, Paris, France, pp 1015–1018
21. Adigbli P (2007) Nonlinear attitude and position control of a micro quadrotor using sliding mode and backstepping techniques. MAV07 & EMAV2007, Toulouse, France, 17–21 September, 2007, pp 1–9

MIX
Papier aus verantwortungsvollen Quellen
Paper from responsible sources
FSC® C105338

If you have any concerns about our products, you can contact us on
ProductSafety@springernature.com

In case Publisher is established outside the EU, the EU authorized representative is:
Springer Nature Customer Service Center GmbH
Europaplatz 3, 69115 Heidelberg, Germany

Printed by Libri Plureos GmbH
in Hamburg, Germany